Wael Bahia
Sarra Souayah
Salima Ferchichi

Expression of miR-21-5p in miscarriage: inflammatory role

Wael Bahia
Sarra Souayah
Salima Ferchichi

Expression of miR-21-5p in miscarriage: inflammatory role

in patients with recurrent spontaneous abortions

ScienciaScripts

Imprint
Any brand names and product names mentioned in this book are subject to trademark, brand or patent protection and are trademarks or registered trademarks of their respective holders. The use of brand names, product names, common names, trade names, product descriptions etc. even without a particular marking in this work is in no way to be construed to mean that such names may be regarded as unrestricted in respect of trademark and brand protection legislation and could thus be used by anyone.

Cover image: www.ingimage.com

This book is a translation from the original published under ISBN 978-620-6-70662-5.

Publisher:
Sciencia Scripts
is a trademark of
Dodo Books Indian Ocean Ltd. and OmniScriptum S.R.L publishing group

120 High Road, East Finchley, London, N2 9ED, United Kingdom
Str. Armeneasca 28/1, office 1, Chisinau MD-2012, Republic of Moldova, Europe
Printed at: see last page
ISBN: 978-620-8-15871-2

Contents

Resume

Recurrent spontaneous abortion (RSA) is defined as two or more consecutive recurrent fatal losses during the first trimesters of pregnancy. Currently, 50% of cases remain unexplained. RSA is a heterogeneous disease, and understanding its pathophysiological mechanisms is a subject of great interest for studies of genetic and epigenetic associations. The inflammation pathway, influenced by hsa-miR-21- 5p and the *NFκBI* and *IL-6* genes, plays a crucial role in maintaining the balance required for a successful pregnancy.

Our master's work consisted in studying the association between the risk of RSA and the three polymorphisms (rs4648143, rs41275743 and rs28362491) of the *NFκBI* gene in a case-control cohort of 243 patients and 251 controls from the Tunisian population. Genotyping was determined by PCR-RFLP and data entry and analysis were performed using SPSS 24.0, Haploview 4.2 and SNPstats.

The results showed an association of the three polymorphisms with the risk of RSA, particularly rs28362491, with a statistically significant difference in allelic and genotypic frequencies between the two study groups. The haplotypic study showed the involvement of 4 haplotypes in the occurrence of RSA, in particular TAG and AGG. As for the study of the differential expression of hsa-miR-21-5p with *NFkB1* and *IL-6* by qPCR, our results do not allow definitive conclusions to be drawn, due to the limited number of samples and the rarity of cases of women who have undergone four or more successive spontaneous abortions without having children.

Key words: ASR, *NFkB1*, *IL-6*, Genetic polymorphism, microRNA, qPCR, RFLP.

Introduction

Spontaneous a repeat abortion (SRA) represents a relatively common reproductive obstacle affecting 15 a 25% of pregnancies (El Hachem et al., 2017). RSA is one of the most complex reproductive anomalies and is frustrating for the couple, their families and treating physicians. Diagnosis and therapeutic intervention rely on knowledge of the risk of fret loss and the proЬaЫШё of finding a treatable ëtiology (Ford and Schust, 2009).

RSA is a multifactorial and poorly understood complication; various factors contribute to its pathogenesis including anatomical abnormalities, endocrine abnormalities, environmental factors, genetic abnormalities and immunological factors. However, the etiology is unknown in approximately 50% of RSA cases, designed as unexplained RSA or idiopathic RSA (IRSA) (Miyaji et al., 2019).

Inflammation, as the body's immune response to pathogens, is considered a major cause of RSA contributing to reproductive failure, including early and recurrent pregnancy loss, due to the deregulation of its mediators including NK cells and T lymphocytes (Kwak-Kim et al., 2014).

Among the various inflammation pathways, the transcriptional factor κB (NF-κB) pathway plays a crucial role in pregnancy (Gomez-Chavez et al., 2021). NF-κB represents a family of five members (NFkB1, NFkB2, RelA, RelB and c-Rel) that regulates innate and adaptive immune responses during pregnancy. It controls the cytokines required for the implantation, invasion and diffërentiation of trophoblastic cells, while maintaining the balance between Th1 and Th2 cytokines to ensure that pregnancy runs smoothly. In contrast, any dysregulation of NF-κB leads to pregnancy loss (Gomez-Chavez et al., 2021; Yang et al., 2022).

NFκBI is a gene encoding two subunits (p50 and p105) of NF-κB, located on chromosome 4 at locus 4q24 and is essential in determining inflammatory responses. A dërëgulation of NF-κB expression and activity could be due to gënëtic variations or single nucleotide polymorphisms (SNPs) of the *NFκBI* gene, which is probably associated with ASRI. Among these polymorphisms, insertion/deletion type rs28362491 is linked to a high risk of unexplained RSA (Arisawa et al., 2013; Misra et al., 2016).

The complexity of the inflammatory response requires a sophisticated regulatory network to perform functions at the level of the signals and genes of the inflammatory process particularly those of the transcription factor of THE NF-κB family. This network consists of a set of ëpigënëtic phënomënes such as тёЛу1айпо 1 DNA and histone modifications that have been shown to be crucial in the regulation of inflammatory genes (Bayarsaihan 2011).

Over the last few decades, one of the most revolutionary epigenetic processes has been the discovery of small non-coding RNAs called microRNAs (or miRNAs) that regulate gene expression at several levels. They are generally known to bind to specific sequences located at the 3' untranslated end (3'UTR) of their target mRNAs, resulting in either translational repression or degradation of the target mRNAs. MiRNAs also bind to the 5'-UTR end, coding region or promoter region of mRNAs, inhibiting gene expression. Inflammation is mainly regulated by miRNAs, since they modulate the expression of inflammatory cytokines and the growth of inflammatory cells. Several diseases associated with pregnancy, such as preeclampsia (PE) and diabetes, are often characterised by an exacerbated inflammatory response, and miRNAs play a vital role in the development of these complications (Zhang et al., 2021; Das and Rao, 2022).

At present, RSA has neither an effective treatment nor a reliable biomarker for diagnosis and prognosis. Circulating microRNAs have recently been considered as potential biomarkers of pregnancy-associated complications due to their fundamental role in placental development and function by regulating trophoblastic cell differentiation, invasion and proliferation (Jairajpuri et al., 2021; Xu et al., 2022).

Some ëstudies have shown the involvement of microRNAs in the pathogenesis of RSA. For example, the miRNA miR-365 plays a role in recurrent abortion by regulating trophoblast apoptosis as well as miR-27a-3p which participates in this complication by decreasing HLA-G expression. The miR 155-5p inhibits the NF-κB signalling pathway in the ASR to allow the growth of decidual cells responsible for placental development (Au - Farine et al., 2018; Zhang et al., 2021).

An association between another microRNA, hsa-miR-21-5p, and inflammation in the context of RSA was previously identified through bioinformatics analysis by Dr. Bahia Wael and colleagues. This research has rëyë!ë a correlation ëйoкe between hsa-miR-21-5p and *NFκBI*, suggestingë that their interaction influences the regulation of 1 inflammation related to ASR. This connection is characterizedëriséе by an iregative feedback, or this microRNA interacts with *NFκBI*, inducing an altëration of its activ^ (Bahia et al., 2020).

However, our present work presents a significant advanceëe by transforming this knowledge into an in vitro ëstudy. By implementing an extensive literature search and rigorous mëthodology to confirm these important links between hsa- miR-21-5p and the *NFκBI* inflammation pathway and also to provide ënew insights into their role in spontaneous rëpëtës abortions. This in vitro approach will strengthen our understanding of the mechanisms underlying this condition, paving the way for future advances in the diagnosis and treatment of these dëlicate situations.

In our ëtude, we will dëterminate the ancient, gënotypic and haplotypic frequencies as well as the gënëtic variabil^ of the three SNPs of the *NFκBI* gëne (rs41275743, rs4648143, rs28362491 INDEL) and to study the gënëtic association between the risk of RSA and these SNPs in a "castitimontrol" cohort of the Tunisian population.

We will ëalso ëstudy the diffërential expression of hsa-miR-21-5p and a group of its inflammation pathway target gënes (*NFκBI* and *IL-6) in* the two groups of female participants.

CHAPTER 1

Summary Bibliography

1. Repeated spontaneous abortions

1.1 Definition

Repeat spontaneous abortion (RSA) or miscarriage is a multifactorial complication that constitutes a real reproductive obstacle for couples trying to conceive, affecting 2 a 5% of them. In addition, 15 a 20% of pregnancies may end in spontaneous loss (Ng et al., 2021). According to a worldwide estimate, 23 million cases occur annually (Turesheva et al., 2023). Differing medical perspectives on RSA lead to a variety of definitions. According to the World Health Organisation (WHO), it is the expulsion of an embryo or fetus weighing less than 500 grams (Sejourne et al., 2008). The American Society of Reproductive Medicine (ASRM) and the European Society of Human Reproduction and Embryology (ESHRE) share a common definition of RSA, although the latter excludes extrauterine pregnancy and molar pregnancy. They both consider RSA as the occurrence of two or more clinical pregnancy losses (ASRM, 2012; ESHRE, 2017). Some literature sources define RSA as the occurrence of two or more consecutive pregnancy losses before 20eme weeks gestation (Dimitriadis et al., 2020; Vomstein et al., 2021).

1.2 Classification

ASRs can be subdivided into two classes:

- **Early ASR**

An early miscarriage is the spontaneous expulsion of an intrauterine pregnancy before 14 weeks' amenorrhoea (SA). Recurrent early miscarriage is defined as three or more consecutive early miscarriages before 14 SA (Delabaere et al., 2014).

- **Late ASR**

A late miscarriage is the spontaneous evacuation of a non-viable fetus between 14 SA and 22 SA or weighing less than 500g (Delabaere et al., 2014; Toupet et al., 2015).

1.3 Causes

Despite the identification of numerous risk factors, RSA is frequently considered as idiopathic RSA, because in 50% of cases, its cause is imprecise or unknown (Jairajpuri et al., 2021).

1.3.1 Anatomical anomalies

Congenital uterine malformations such as arcuate, septate, unicornuate, bicornuate and didelphic uteri (Bashiri, 2016; Dimitriadis et al., 2020), as well as acquired malformations such as endomëtrial polyps, intra uterine adhërences and uterine fibroids can be associated with ASR during the 1st and 2eme trimesters of pregnancy (Turocy and Rackow, 2019).

1.3.2 Thrombophilia

Thrombophilia is an abnormality that affects blood coagulation, increasing the risk of venous or arterial thromboembolic diseases. It can be hereditary, linked to genetic mutations in factor V Leiden and factor II, or acquired, associated with anti-phospholipid antibodies (aPL) linked to pregnancy loss during the 1er and 2eme trimesters (ESHRE, 2017; Dimitriadis et al., 2020).

1.3.3 Endocrine factors

I.3.3.1 Thyroid function

Disturbances in thyroid hormones and anti-thyroid immunity are responsible for disturbances in fertilisation, eenbrvogenesis and therefore pregnancy loss (Dimitriadis et al., 2020). Anti-thyroglobulin (anti-TG) and anti-thyroperoxidase (anti-TPO) antibodies significantly increase the risk of miscarriage (Dong et al., 2020; Godines-Enriquez et al., 2021).

I.3.3.2 Polycystic ovary syndrome (PCOS)

This syndrome has been associated with numerous pregnancy complications such as fret loss and gestational diabëte. However, there is insufficient evidence to state that PCOS is an independent risk factor for RSA (Dimitriadis et al., 2020).

Patients with PCOS have a significant prevalence of thrombophilia, leading to more ëlevës rates of pregnancy loss (Kazerooni et al., 2013).

1.3.4 Infectious diseases

Infections with Cytomëgaloviшs, Herpes simplex virus, rubëole, toxoplasma, syphilis and mycoplasma can contribute to RSA. Infection with *Chlamydia trachomatis* is strongly associated with various dëfavoraЬles outcomes of pregnancy such as RSA as well as bacral vaginosis which appears to increase the risk of this complication (Merviel et al., 2005; Shaheen and Saxena, 2022).

1.3.5 Environmental factors

1.3.5.1 Obesite

Obesity represents an indëpendant risk factor for 1RSA due to its impact on women's reproductive healthë. It is дёГтс, according to the WHO, by a body mass index (BMI) > 30Kg/m^2 and is associated with subfertil^ as well as an increased risk of pregnancy loss (ESHRE, 2017; Dimitriadis et al., 2020).

1.3.5.2 Maternal age

The advanced ageë of the woman is a well ë-established factor in recurrent pregnancy loss as it is implicated in female subfertility and fretal anomalies. It is a determining factor in the prognosis of a live birth (ESHRE, 2017). The risk of RSA is 9 to 12% in women under 35. In contrast, it increases up to 50% in women over 40 compared to younger women (El Hachem et al., 2017).

1.3.5.3 Smoking

Maternal smoking is known to have adverse effects on reproduction, such as increased risks of placental dëcollement, bleeding during pregnancy and reduced fertility due to nicotine restricting blood flow to Harris and the placenta (Bashiri, 2016). № Nevertheless, there is controversy regarding the association between smoking and pregnancy loss, as some ëtudies have not found6 a significant correlation (Ford and Schust, 2009; Ng et al., 2021).

1.3.5.4 Alcoholism

Alcohol consumption has a negative impact on pregnancy as it easily crosses the placental barrier. An amount of alcohol in excess of 80g/week leads to Fetal Alcohol Syndrome. Consequently, it could be responsible for intrauterine growth retardation and fretal growth arrest (Benammar et al., 2012).

1.3.5.5 Consumption of caffeine

The cafëine consumed from cafë, Лё and certain soft drinks crosses the placental barrier and reduces placental blood flow by increasing levels of circulating catëcholamines. Cafëine appears to be associated with ASR as it may increase the risk of abnormal fretal cardiovascular development and intra -^rine growth retardation (Benammar et al., 2012).

1.3.6 Male causes

The integrity of the sperm genome is crucial to the success of a pregnancy. Several male factors such as sperm quanta and quality^, gënëtic mutations or polymorphisms, sperm chromosome structure and number abnormalities, sperm DNA fragmentation as well as epigenetic changes and age of the përe have received significant attention, as they have been associated with ASR. (Yu and Bao, 2022).

1.3.7 Genetic factors

1.3.7.1 Chromosomal abnormalities

Chromosomal abnormalities are responsible for 50-60% of early ASR and may be of parental origin or de novo. The most common parental anomalies include reciprocal translocations, Robertsonian translocations and inversions (El Hachem et al., 2017; Priya et al., 2018).

Fetal chromosomal aberrations include structural anomalies, gënëtic mosaicism, and aneuploi'die. The latter is consideredërëe the most fequent cause of early RSA, the rate and complexity of which increases significantly with maternal age (Bashiri, 2016; Kaser, 2018).

1.3.7.2 Genetic mutations

Numerous mutations are associated with RSA, notably those affecting factor V Leiden and the prothrombin gene, as well as polymorphisms in genes related to vascular endothelial growth factor (VEGF), endothelial nitric oxide synthase (eNOS), matrix metalloproteinases (MMPs) and progesterone and estrogen receptors (Mtiraoui et al, 2005; Bahia et al., 2020).

1.3.8 Immunological factors

1.3.8.1 The HLA system

The HLA system could play a crucial role in fertility, pregnancy establishment and embryonic survival. The compatibility of the mother's immune system with that of her fetus is crucial to the maintenance of pregnancy, given that the mother and fetus are genetically different. HLA-G, expressed on the surface of trophoblast cells, is one of the major molecules involved in regulating the maternal immune response during pregnancy. Alterations in the HLA-G pathway contribute to immunological imbalance and increase the risk of miscarriage (Yazdani et al., 2018; Barbaro et al., 2023).

1.3.8.2 NK cells

Uterine regulatory NK cells (uNK) are highly granular NK cells found in the mucosa of the uterus. During a normal pregnancy, an increase in the number of these cells is essential to ensure reproductive success, as they play a fundamental role in trophoblast invasion and adequate placentation. Indeed, a decrease in the number of these uNK cells or their dysfunction is closely associated with RSA (Vomstein et al., 2021; Yang et al., 2023).

1.3.8.3 T lymphocytes

T helper lymphocytes (THLs) play a central role in modulating immune responses. During pregnancy, each subset of Th cells, such as Th1, Th2, regulatory T cells (Treg) and Th17, performs variable functions including fretal development and immunological surveillance (Graham et al., 2021).

Treg play a dominant role in the maintenance and immune toterance of the fretus during pregnancy, ëlevës levels of Treg are dëtectës in normal pregnancy while a reduced population of CD4+ Treg a ëlë signage in women with RSA (Zhao et al., 2020; Wang et al., 2022).

Th17 cells, which produce the inflammatory cytokine interleukin-17 (IL-17), provide defence against infections, autoimmune diseases and inflammatory disorders. This lineage influences the success of pregnancy as well as the development of pregnancy-related pathologies such as ASR. Th1 and Th2 cells sëcrëtent cytokines to ensure the smooth running of the various stages of pregnancy, in particular implantation and placentation. However, an imbalance in their activities can lead to spontaneous abortion (Fu et al., 2014; Graham et al., 2021).

1.3.8.4 Cytokines

Cytokines are proteins produced by immune cells, particularly THLs, and play a crucial role in cell development, communication and the regulation of the immune response (Zhang et An, 2007; Chauhan et al., 2021).

There are two main types of cytokines that are produced by different populations of LTh cells, Th1-type cytokines called pro-inflammatory cytokines (IL-ie, IL-2, IL-6, IL- 12, IL-18, TNF-a, IFN-y) that amplify the inflammatory response and Th2-type cytokines (IL-4, IL- 5, IL-10, IFNa, TGFe), nommëes anti-inflammatory cytokines, that repress this response (Lee et al., 2019; Lordan et al,. 2019).

Certain cytokines can have a positive or negative impact on pregnancy. For example, cytokines such as IL-4, IL-6 and IL-10 seem to favour a successful pregnancy, whereas

1 IFNY and TNFa can be detrimental by altering the growth and function of trophoblast cells (Daher et al., 2004; Bashiri, 2016).

11. Inflammation

11.1 Description

Inflammation is an essential defence mechanism for health, and is a biological response of the immune system aimed at eliminating harmful stimuli such as pathogenic agents, damaged cells, toxic compounds and irritants, which trigger inflammatory responses in several vital systems, including the reproductive system, by activating inflammatory cells and inflammatory signalling pathways which are most often the NF-κB, MAPK and JAK-STAT pathways. This inflammatory response thus promotes tissue repair and the healing process (Chen et al., 2018).

In addition, inflammatory immune responses are a major cause of RSA involved in reproductive failure, including early and recurrent pregnancy loss. Cellular immune responses, particularly those mëdiëed by NK cells and T cells, are most often dysregulated in this context (Kwak-Kim et al., 2014). It is ëgalement emphasised that prolonged and uncontrolled inflammatory responses can impair fetal growth and development. Ultimately, inflammation, whether systemic or local, is closely associated with RSA, suggesting the involvement of the inflammatory process in RSA (Jiang et al., 2021).

11.2 Mechanism of inflammation

11.2.1 During a normal pregnancy

Pregnancy is both a pro-inflammatory and anti-inflammatory condition, depending on the stage of gestation.

During implantation, placentation and the first and second trimesters of pregnancy, a strong Th1-type pro-inflammatory response is necessary to ensure the adequate reparation of the uterine epithelium and the elimination of cell damage.

The second immunological phase of pregnancy is a period of fetal growth and development (**Figure 1**). The mother, placenta and fetus are symbiotic, establishing a Th2-type anti-inflammatory state. Finally, childbirth takes place through renewed inflammation. This pro-inflammatory condition promotes contraction of the uterus, expulsion of the baby and rejection of the placenta (Mor et al., 2011).

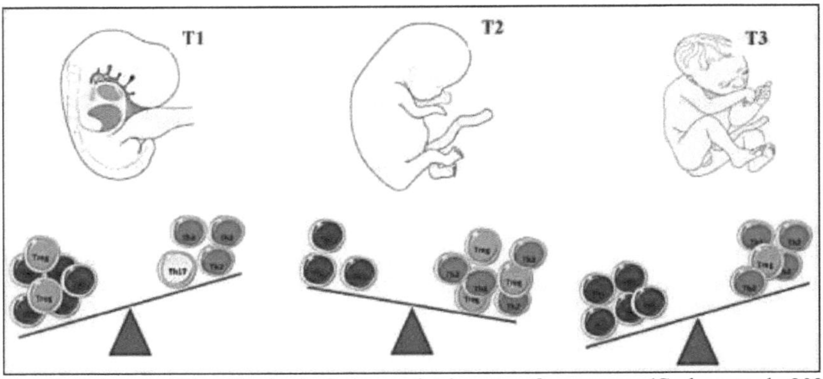

Figure 1: Balance of Th cell subsets during each trimester of pregnancy (Graham et al., 2021)

<u>Description</u>: The first trimester of pregnancy (**T1**) is associated with a predominantly Th1 cell environment necessary for blastocyst implantation. After implantation, the second trimester (**T2**) is characterised by a Th2 anti-inflammatory immune environment, necessary for fetal development. Finally, during the third trimester (**T3**), there is a shift to an inflammatory immune state associated with Th1 cells, necessary for parturition. Treg cells are present in every trimester because of their role in the success of pregnancy, by maintaining maternal immune tolerance to her fetus.

11.2.2 In the ASR

A dërëglement of Th cells as well as overactivation of NK cells are associated with multiple obstetric complications and even RSA (Graham et al., 2021; Jiang et al., 2021).

A disequilibrium not only between Th1/Th2 cells, marked by a remarkable predominance of Th1 over Th2 immunity favouring an unfavourable implantation environment, but also a dysregulation between Th17/Treg cells seems to lead to RSA (**Figure 2, Figure 3**) (Park et al., 2022).

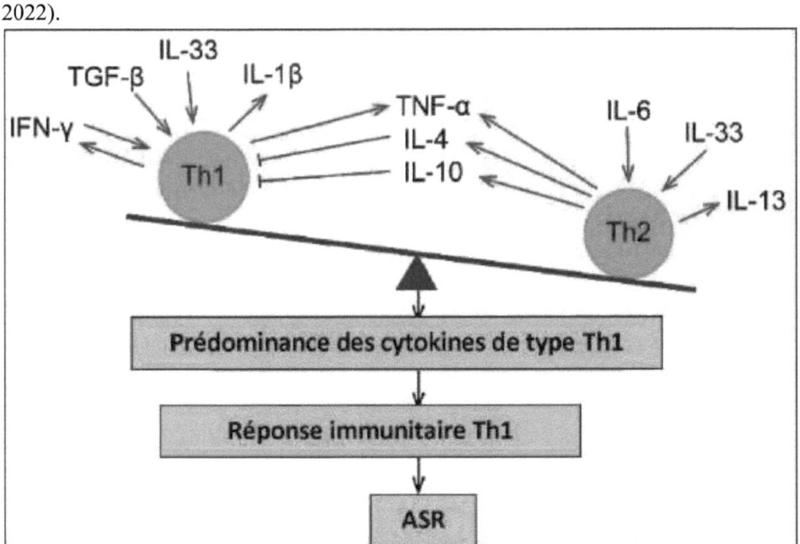

Figure 2: Th1/Th2 cytokine imbalance in patients with RSA (Yang et al., 2023)

Description: i'JFN-y, secreted by Th1 cells, in association with TGF-β and IL-33, ensures the differentiation of Th1 cells. IL-33 and IL-6 promote the differentiation and stimulation of Th2 cells which produce IL-4 and IL-10, inhibiting the production of Th1 cells. Deregulation of the ratio of Th1 to Th2 cells triggers a pro-inflammatory Th1 response, favouring pregnancy failure and leading to RSA.

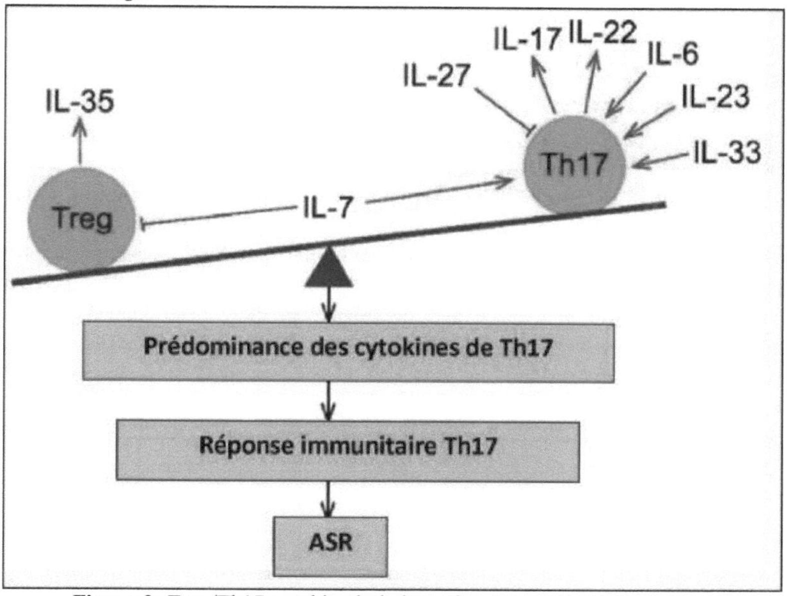

Figure 3: Treg/Th17 cytokine imbalance in ASR patients (Yang et al. 2023)

Description: IL-7 promotes the differentiation of Th17 cells and inhibits Treg cells, thereby compromising immune tolerance and promoting inflammatory responses. IL-33 contributes to the immune response by Th17 cells and ensures their differentiation, as does IL-6, while IL-23 increases their number. IL-27, on the other hand, inhibits the expression of Th17 cells. Consequently, abnormal production of these cytokines in the peripheral blood can lead to an immune imbalance between Treg and Th17 cells, triggering a pro-inflammatory immune response that can eventually lead to RSA.

11.3 Inflammation pathway

II.3.1 The NF-κB pathway

The transcriptional factor nucieaire κB (NF-κB) is a nuclear factor involved in various physiological and pathological processes of the immune and inflammatory responses and which controls the transcription of more than 400 genes, including those of cytokines (Sun, 2017; Gomez-Chavez et al., 2021).

This family is composed of five structurally related members, including NFκBI (also called p50 and its precursor p105), NFκB2 (also called p52 and its precursor p100), RelA (or P65), RelB and c-Rel, which mediate transcription of target genes by binding to a specific DNA element, the κB enhancer, in the form of heteros or homodimers (Baud and Jacque, 2008; Liu et al., 2017).

The transcriptional activity of these proteins is regulated by two main pathways:

- **The canonical NF-κB: activation pathway** is triggered by various stimuli from several receptors such as cytokine receptors, and applies predominantly to RelA/p50 dimers. In the absence of activation, these dimers are sequestered in the cytoplasm by their association with members of the κB inhibitor (IκB) family, in particular 1κBa (Baud and Jacque, 2008; Sun, 2017).
- **The non-canonical or alternative activation pathway**: is induced by different members of the TNF receptor superfamily (TNFR) such as CD40L and BAFF (*B cell-activating factor)*. This pathway applies mainly to RelB. It is triggered by different members of the TNF (*tumour necrosis factor)* superfamily and relies on induced proteolysis of the p100 protein, the main inhibitor of RelB. In the absence of activation, the RelB/p50 and RelB/p52 dimers are sequestered in the cytoplasm via their association with p100 (Baud and Jacque, 2008; Sun, 2017).

Activation of the canonical pathway, illustrated in **Figure 4**, occurs in response to a number of stimuli, including cytokines. The trimeric IκB kinase complex (IKK), composed of the catalytic subunits IKKa and IKKe and the regulatory IKK (also known as NEMO), phosphorylates IκB proteins, notably 1κBa, which are then ubiquitinylated and finally degraded by the 26S proteasome. The RelA/p50 dimers thus freed from their inhibitor then reach the nucleus where they bind to specific DNA sequences known as the κB enhancers of the target genes and thus activate the transcription of these genes (Baud and Jacque, 2008; Sun, 2017).

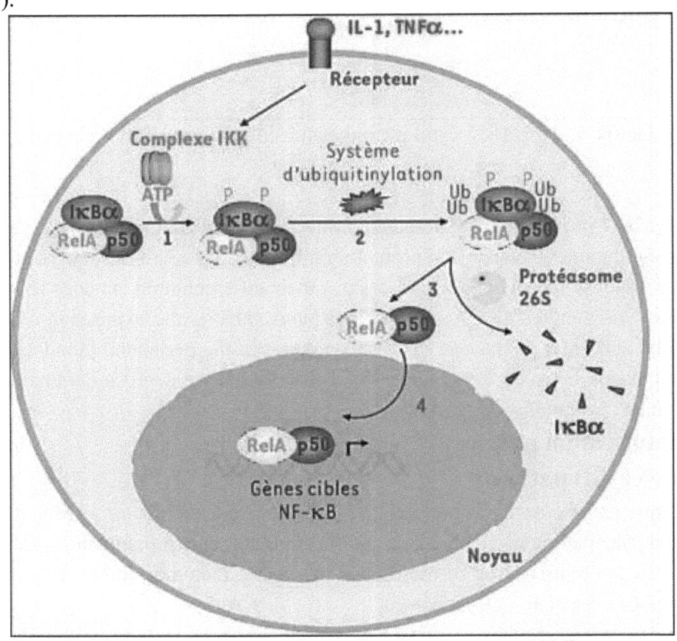

Figure 4: Canonical NF-κB activation pathway (Baud and Jacque, 2008)

Description : Upon stimulation of the canonical pathway by TNFa or IL-1, the IKK complex induces phosphorylation of the inhibitory protëines IκB **(1)**. Subsequently, IκBs undergo ubiquitinylation **(2)** and dëcomposëes by the 26S protëasome. This allows the Hbëraйoп of RelA/p50 dimëres **(3)**, which migrate to the nucleus to activate transcription of their target gënes **(4)**.

In response to one of the stimuli by one of the TNF protëins, the non-canonical pathway is activated (**Figure 5**). The protëine NIK (*NF-kB-inducing kinase)*, associated in the cytoplasm with the adaptor protëine TRAF3 *(TNF receptor-associated factor 3)*, is directed towards the plasma membrane where TRAF3 is then degraded. NIK released in this way autophosphorylates, phosphorylates and activates IKKa, which in turn phosphorylates p100 associated with RelB, resulting either in the total dëgradation of p100 and the release of RelB/p50 dimëers, or in the partial protëolysis of p100. In consëquence, this releases RelB/p52 dimëers which, in turn, are responsible for the transcription of target genes (Baud and Jacque, 2008; Sun, 2017).

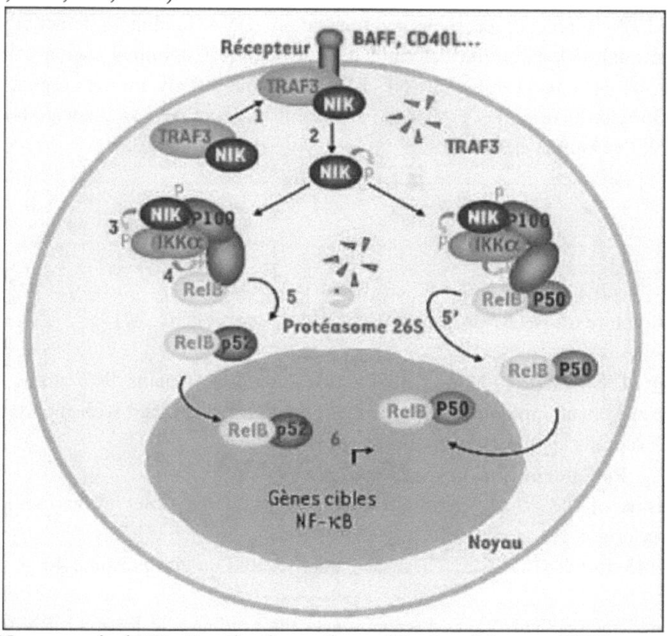

Figure 5: Non-canonical NF-κB activation pathway (Baud and Jacque, 2008)

Description : When receptors of the TNFR family, such as BAFF and CD40L, are active, the TRAF3-NIK complex is recruited to the membrane **(1)**. Subsequently, TRAF3 undergoes dëgradation and NIK is activated by autophosphorylation **(2)**. Formation of the NIK- IKKa-p100 triple complex **(3)** leads to phosphorylation of p100 **(4)** and its ubiquitinylation. This mechanism can lead either to the maturation of p100 to p52, generating RelB/p52 dimers **(5)**, or to the degradation of p100, releasing RelB/p50 dimers **(5')**. These dimers can then activate transcription of their target genes **(6)**.

11.3.1.1 The *NFκBI* gene

The human *NFκBI* gene is located on chromosome 4 at locus 4q24. It is 156 kb long and comprises 24 exons, with introns ranging in length from 40,000 to 323 bp. This gene encodes a protein of the Rel family, called p105, composed of 968 amino acids, with a molecular weight of 105 kDa. This protein acts as a transcription inhibitor and undergoes post-translational cleavage by dëgradation of the C-terminal part of *NFκBI,* dëpending on the 26S protëasome. This cleavage produces a 50 kDa protëine (p50) that represents the DNA-binding subunit of the NF-κB protëic complex and corresponds to the N-terminus of p105 (Karban et al., 2004; Edenberg et al., 2008; Misra et al., 2016).

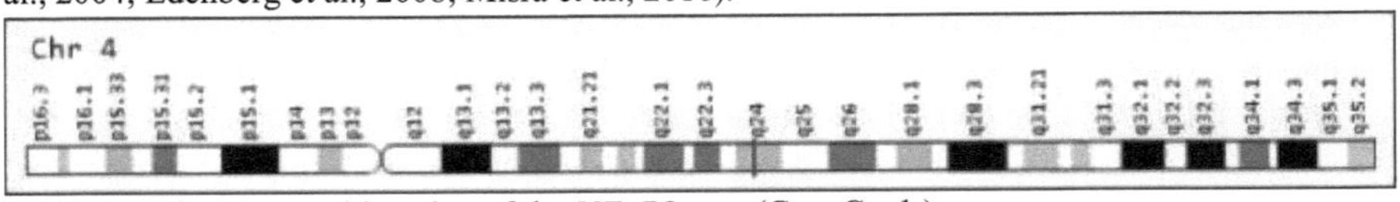

Figure 6: Chromosomal location of the *NFκBI* gene (GeneCards)

Each of the *NFκBI* protëic subunits contains the Rel homology domain (RHD) near its N-terminus, which is crucial for subunit dimerisation, DNA binding, interactions with IκB proteins and nuclear localisation determined by a flexible C-terminal region containing the nuclear localisation signal (NLS). The NF- κB p50 subunit has glycine-rich regions (GRRs) in the 375-400 region that are necessary and sufficient to direct *NFκBI* cleavage (Huxford and Ghosh, 2009; Peterson et al., 2011).

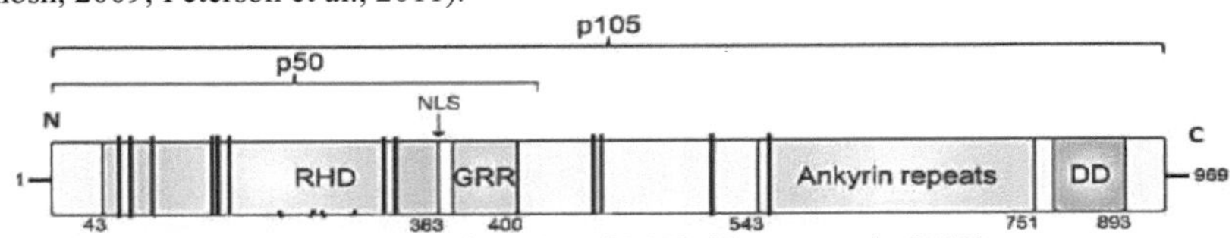

Figure 7: Structure of *NFkB1* protein P50 and P150 (Hoeger et al., 2017)

Description: The *NFκBI* protein, with its two subunits, contains 969 amino acids. The Ankyrin repeats enable protein-protein interaction and are combined with the *Death Domain* (DD), which recruits apoptosis-promoting proteins.

11.3.1.2 Polymorphisms in the *NFκBI* gene

Polymorphisms of the *NFκBI* gene are associated with the onset of various diseases of inflammatory origin, for example :

- **rs4648143 and rs41275743:** These SNPs are located on chromosome 4q24, precisely in the

3' untranslated region (3'UTR), corresponding to the binding site for microRNAs, the main regulators of the inflammatory process (Sun et al., 2020).

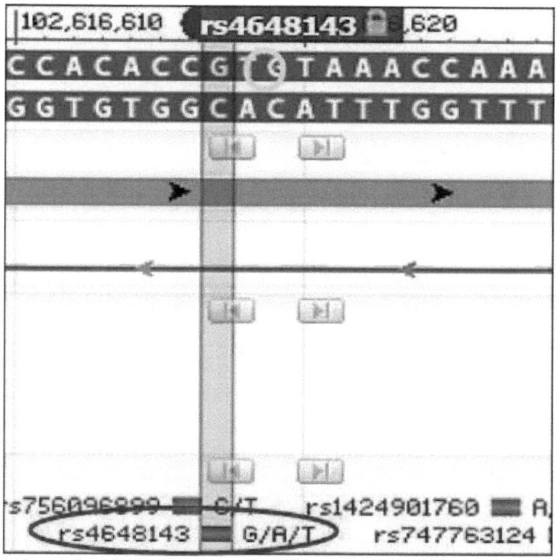

Figure 8: Localisation of rs4648143 of the *NFkB1* gene (GRch38, 2022)

Description: Variation of the G allele (circled in yellow) on the 5'-3' template strand can result in either a transition to Adenine (A) or a transversion to Thymine (T), the latter being extremely rare.

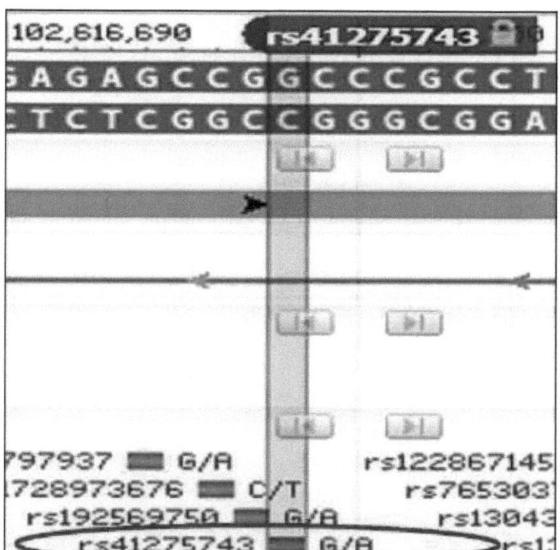

Figure 9: Location of rs41275743 of the *NFκBI* gene (GRch38, 2022)

- **rs28362491 INDEL**: This insertion/deletion polymorphism of the ATTG nucleotide, situc

in the promoter of the *NFκBI* gene at locus 94 (-94ins/del ATTG), influences the expression of the P50 protëine and modifies the level of inflammatory responses. The presence of a 4 base pair (bp) dëlëtion results in reduced promoter activity. This SNP is ëtropically associated with infertility Hëc a Гendomëtriosis and idiopathic infertility (Bianco et al., 2012; Sun et al., 2020).

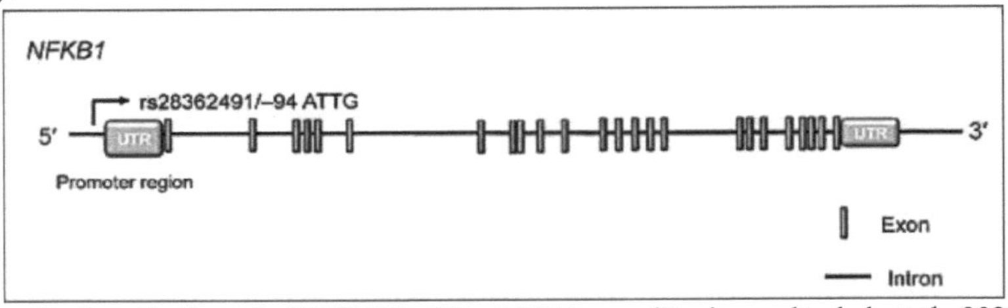

Figure 10: Localisation of rs28362491 of the *NFkB1* gene (Kanigur sultuybek et al., 2020)

11.3.1.3 Pathophysiology of THE NF-κB pathway

NF-κB regulation is crucial in pregnancy, playing an essential role in the innate immune response of the female genital tract prior to gestation. A reduction in its activity may favour infections, while exagërëe activity may cause excessive inflammation. Cytokines controlled by NF-κB are essential for fetal implantation by promoting trophoblast invasion and differentiation. However, excessive activation of NF-κB after implantation can cause impaired fetal development, potentially leading to premature birth or abortion. During pregnancy, NF-κB is negatively regulated in maternal T cells, establishing a balance between Thl and Th2 cytokines, with a predominance of Th2 cytokines that is reversed at the end of gestation. When this balance is abnormal, it can initiate a cascade of production of inflammatory cytokines harmful to pregnancy (Challis et al., 2009; Gomez-Chavez et al., 2021).

The intracellular signalling cascades that regulate inflammation are subject to many levels of regulation, including regulation by an epigenetic process. Epigenetics is defined as heritable changes in gene expression that, unlike mutations, do not alter the DNA sequence. The predominant epigenetic mechanisms are DNA methylation, chromatin modifications and regulation by non-coding RNAs (Hamilton, 2011).

Epigenetic mechanisms are well known to be involved in the onset and development of recurrent pregnancy loss by regulating the expression of key genes determining cell differentiation, proliferation and apoptosis (Hocher and Hocher, 2018; Zhou et al., 2021).

111. Epigenetic factors

111.1 DNA methylation

DNA methylation is one of the main epigenetic mechanisms that repress gene expression and is involved in regulating the other two mechanisms, namely histone modification and regulation of gene expression by non-coding RNAs. Methylation takes place at cytosine residues at position C5 in CpG dinucleotides located in the promoter region of the gene. CpG sites subject to methylation are distributed unevenly throughout the genome, forming groups known as CpG islands (Kiselev et al., 2021).

Methylation is involved in implantation and embryonic development. Abnormal DNA methylation is a potential cause of early pregnancy loss, associated with abnormal embryonic development and miscarriage. Worldwide, around 4% of spontaneous abortions have been linked to abnormal DNA methylation (Pei et al., 2019; Zhou et al., 2021).

111.2 Non-coding RNA

111.2.1 Long non-coding RNA

Long non-coding RNAs (lncRNAs) are defined as repeat-rich non-coding transcripts of more than 200 nucleotides. They are mainly transcribed by RNA polymerase II and generally range in size from around 1 kb to over 100 kb. These RNAs are actively expressed during cell differentiation and development, and regulate several physiological processes such as cytokine expression and inflammation (Statello et al., 2021; Mattick et al., 2023).

LncRNAs are among the epigenetic factors of interest in the regulation of gene expression, and research on these RNAs could contribute to the understanding of the mechanisms involved in idiopathic RSA (Arias-Sosa et al., 2018).

111.2.2 Small RNA interferent

Small interfering RNAs (siRNAs or RNAi) are small non-coding RNAs of 20 to 30 nucleotides capable of regulating the expression of a large number of genes by a biological mechanism known as RNA interference, which induces gene silencing by targeting the complementary mRNA in order to degrade it (Dana et al., 2017; Fattal, 2020).

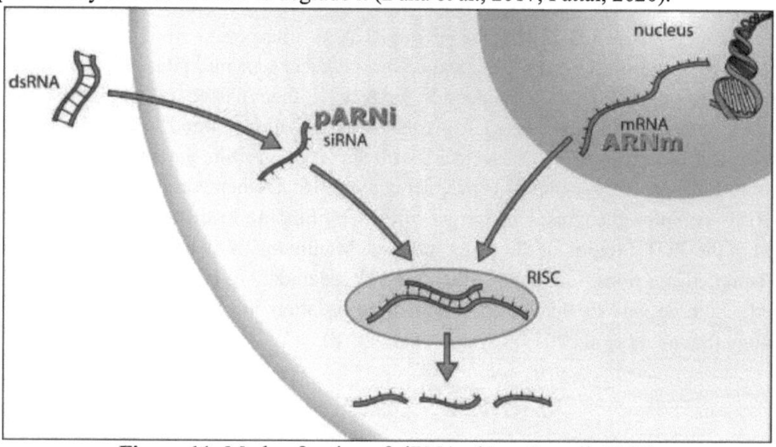

Figure 11: Mode of action of siRNAs (AquaPortail, 2021)

Description : Extracellular double-stranded RNA (dsRNA) (from certain viruses) is cleaved into siRNA (or pRNAi for small interfering RNA) by the action of an RNase III-type enzyme called Dicer. The siRNAs thus formed are in the form of dsRNAs, 21 to 23 bp long, which bind to a cytoplasmic protëic complex called the RISC complex (*RNA induced silencing complex*) in which one of the two strands is lilar while the other, whose 5' end is the more stable, will be integrated into the RISC complex and associate with the argonaute protëine Ago2 which takes it to its complementary target sequence on the mRNA. The target mRNA is finally cleaved through the catalytic activity of the RISC complex, inhibiting its translation into protëine (Dana et al. 2017; Fattal 2020).

111.2.3 MicroRNA

III.2.3.1Generates

MicroRNAs (miRNAs) are small single-stranded non-coding RNAs of around 22 nucleotides involved in the post-transcriptional regulation of around 60% of genes by binding to the target mRNA and affecting its translation. MiRNAs are involved in almost all physiological and pathological processes in the body, including proliferation, growth, development, cell differentiation and apoptosis (Chen et al., 2021; Jairajpuri et al., 2021). There are around 2000 miRNAs in the human genome, located either in intergenic regions or in the transcriptional regulators of target genes (Jie et al., 2021).

MiRNAs are involved in the regulation of signalling pathways, including NF-kB, and have a negative impact on gene expression during implantation and embryonic development (Pei et al., 2019; Zhu et al., 2021). For this reason, several studies have highlighted the importance of miRNA profiling and expression analysis in the context of RSA as reliable biomarkers for the diagnosis and prognosis of patients with this complication (Bahia et al., 2020; Jairajpuri et al., 2021).

111.2.3.1 Mode of action of microRNAs

The genes coding for miRNAs are transcribed by RNA polymërase II into pri-miRNAs which are cleaved and transformed in the nucleus into simple hairpins called prScursors (prS-miRNA) of around 70 nucleotides, by a nuclear protSic complex called a microprocessor consisting of the enzymes Drosha, DGCR8 (*Di George critical region 8*) and double-stranded RNA-binding protëine (dsRBP). This prS-miRNA is transported from the nucleus to the cytoplasm by Exportin 5 where it is cleaved by the Dicer enzyme, releasing a small double-stranded miRNA of 20 to 25 nucleotides, which is then passed on to one of the four Argonaute protSine families (Ago 1 to 4) that selects a single strand to become the mature miRNA. The mature miRNA, associated with the Ago protSin, interacts with the TRBP protSin to form the RISC complex (Savagner et al., 2015; Treiber et al., 2019).

This RISC complex recognises the target mRNA by binding to a site of 2 to 8 nucleotides located in the 3'UTR region of the target mRNAs. Mature miRNAs hybridise to this region by base complementarities, inducing either mRNA degradation if the miRNA sequence is perfectly homologous to that of its target, or translation inhibition in the case of partial homology (Savagner et al., 2015; O'Brien et al., 2018).

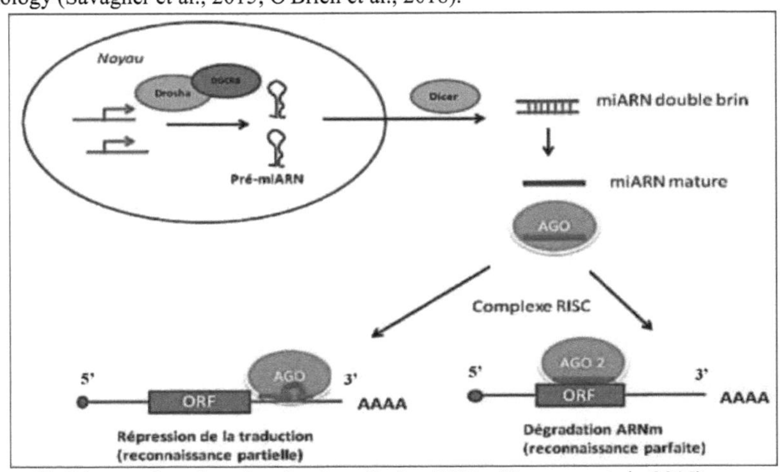

Figure 12: Biogenesis and mode of action of microRNAs (Savagner et al., 2015).

Description : The Ago2 protein of the RISC complex binds to the reading frame (ORF) of the target gene and causes dëgradation of the mRNA, while another Ago protëin binds close to the ORF of the gene in question, inhibiting its translation into protein.

111.2.3.2 Role of miRNAs in normal pregnancy

miRNAs contribute to the development and maintenance of pregnancy by regulating important stages such as trophoblast differentiation and migration, embryo implantation, maternal-fetal immune tolerance and embryonic growth and development. Nevertheless, their role, level of expression and variations during the different stages of pregnancy remain poorly understood overall (Legare et al., 2022).

111.2.3.3 Role of miRNAs in RSA

MicroRNAs are discernible in most fluids such as plasma or serum. Aberrant expression of these miRNAs is responsible for several inflammatory, cardiovascular and gynecological diseases, including RSA (Jairajpuri et al., 2021).

Dysregulation of miRNAs in the placenta has an impact not only on placental development and function, but also on fetus growth. Because of their variation in expression in the placenta and maternal bloodstream during pregnancy complications, miRNAs can be used as diagnostic biomarkers. They play an essential role in regulating genes involved in the proliferation, differentiation, invasion, migration, apoptosis and angiogenesis of trophoblastic cells, thereby influencing placental development. Abnormal expression of miRNAs can disrupt trophoblast function, leading to serious pregnancy complications, including RSA (Ali et al., 2021; Hakimi et al., 2023).

It has been established that miRNAs are involved in the regulation of immune and inflammatory responses, as well as spontaneous miscarriage. Among these miRNAs, hsa-miR-21-5p appears to play a major role in the inflammation pathway in women with RSA (Bahia et al., 2020; Salimi et al., 2022).

111.2.3.4 Hsa-miR-21-5p

Several pathways involved in RSA are regulated by miRNAs, in particular the TGF-β signalling pathway, thyroid hormones and MAPK signalling pathways, which target the *NFκBI* gene. The association between hsa-miR-21-5p and the *NFκBI* gene is involved in the control of inflammation, which is a risk factor in RSA. *NFkB1* activity is regulated by the ubiquitin protein ligase pellino homologue 1, whose expression is negatively regulated by hsa-miR-21-5p, thereby inhibiting *NFkB1* activity (Bahia et al., 2020).

According to the miRBase, RNA central and UCSC Genome Browser databases, hsa-miR-21-5p is a mature 22 bp microRNA with the sequence UAGCUUAUCAGACUGAUGUUGA, encoded by a gene located on chromosome 17q23.1.

Figure 13: Chromosomal location of hsa-miR-21-5p (RNACentral)

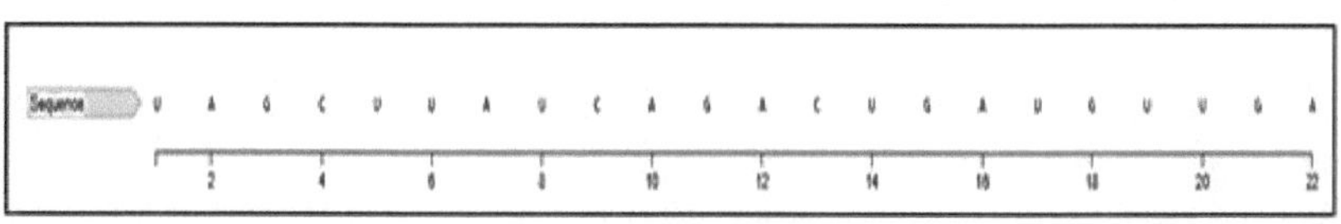

Figure 14: The nucleotide sequence of hsa-miR-21-5p (RNACentral)

IV. Molecular markers of genetic variability in human genetics

Advances in molecular biology have made it possible to identify genetic markers, which are short DNA sequences with a specific location on the chromosome. Molecular marking brings together a set of techniques that rëyëlenl: differences in DNA sëquences between individuals. These techniques can dëtect DNA point mutations such as AFLP "*Amplified fragment length polymorphism*" and RFLP "*Restriction fragment length polymorphism*" or tandem repeats of microsatellite-type nucieotide motifs as well as SNP "*Single Nucleotide Polymorphism*" markers which rëyëкШ: diffërences of a single base at the level of a known DNA sëquence.

IV.1 RFLP markers

RFLP, established by Grodzicker and colleagues in 1974, is used to identify DNA polymorphisms between different individuals based on the unique, highly specific recognition property of bacterial restriction enzymes, which geiK'rent fragments of different sizes depending on the presence or absence of a polymorphism. Its principle is based on the digestion of genomic DNA from various individuals into DNA fragments of variable size, using known restriction enzymes. The digested fragments are then separated by electrophoresis. The molecular basis of RFLP lies in the fact that nucleotide base substitutions, insertions, deletions, duplications and inversions throughout the genome can delete or create new restriction sites (Yang et al. 2013).

RFLP has ël e the first DNA-based marker allowing the construction of gënëtic linkage maps and the realisation of an individual gënëtic fingerprint, which makes it possible to distinguish between two individuals (Yang et al. 2013).

IV.2 Microsatellite markers

Microsatellites, also known as sëquences microsatellites, are fragments of DNA in which a sëquence is rëpëtëcutive in a pattern of 1 to 10 nucteotides. These rëpëtitions, also known as *Simple Sequence Repeats* (SSRs), *Short Tandem Repeats* (STRs) or *Variable Number Tandem Repeats* (VNTRs), are found more frequently in the introns of gënes than in the exons. The number of rëpëtitions, varying from around 10 to 100, differs from cell to cell in the same individual due to errors during DNA replication. They are usedës as gënëtic markers for individual identification and gënëtic mapping (Vieira et al., 2016).

In molecular biology, the flanking regions of microsatellites are used as primers in PCR. Even if a microsatellite is not specific to a particular locus, its flanking regions are. Thus, a pair of primers designed to target these flanking regions amplifies only that spëcific microsatellite. After PCR, a ëlectrophorëse on agarose gel sëpare the DNA fragments according to their size, corresponding to the length of the microsatellite sëquence. Because of the ëlevë polymorphism of these microsatellites due to the number of rëpëtitions of the motif, it is possible to distinguish individuals within an initial population and to diffërenciate homozygous and hëtërozygous individuals (Vieira et al. 2016).

IV.3 SNPS

SNPs represent the most frequent form of genetic variation in humans, with variable frequencies in the general population. They are associated with inter-individual genetic and phënotypic variation, differences in susceptibility to disease and differences in sensitivity to treatment. They are germline mutations present in coding and non-coding regions of DNA, such as inte^nic sëquences, 3' or 5' untranslated regions, intronic regions or transcription factor binding sites (Korzeniewski et al., 2013).

A SNP is defined by the presence of at least two alleles with a frequency greater than 1% in a large population of unrelated individuals. Although the number of SNPs is estimatedë at 10-11 million, the vast majority of SNPs have only two alleles due to the rarityë of mutations at a given position (B0rsting and Morling, 2013).

In general, there are two categories of sequence variations, transition-type point polymorphisms such as purine to purine (A or G) or pyrimidine to pyrimidine (C or T), transversion-type (purine to pyrimidine and vice versa) as well as insertions and deletions (INDELs) (Neininger et al., 2019). The main advantage of using SNPs in genetic studies is that they enable genotyping of a large number of individuals at low cost.

V. From the "candidate gene" approach to pangenomic genetic analysis

V.1 "Candidate" gene approach

This approach focuses on polymorphisms in genes encoding known proteins with a well-defined effect at the cellular level. The polymorphisms selected are those located in exonic sequences, which can lead to a variation in the amrë acid of the translated protein, as well as in promoter sequences likely to modify the transcription level of mRNAs. It is also possible to select polymorphisms in the intronic sequences involved in pre-messenger RNA splicing.

V.2 Pangenomic analysis (GWAS)

Genome-wide association *study (*GWAS) involves the analysis of numerous genetic variations in many individuals. These variations include copy number variations (CNVs), sequence variations in the human genome, and more commonly, SNPs. The aim is to study the correlations between these genetic variations and phenotypic traits (Witte, 2010; Uffelmann et al., 2021).

Population size is very important for obtaining significant results on phenotypic and genotypic variation. Thus, a small number of individuals is a disadvantage that reduces the power of the pangenomic association study. Furthermore, only traits with moderate to high heritability estimates should be considered in pangënomic association studies. Low hëritabilityë in the strict sense is a limiting factor that reduces the ability of GWAS to dëtect 1 association (Alqudah et al., 2020).

CHAPTER 2

Materials & Methods

I. Equipment

I.1 Study population

This is a retrospective case-control study conducted at the Clinical and Molecular Biology Research Unit (UR17ES29) of the Faculty of Pharmacy in Monastir. This study involved 494 Tunisian women: 243 patients with ASR and 251 controls with normal pregnancies and no gynaecological or obstetric complications.

Patients were recruited at the Monastir neonatology and maternity centre, and the collection procedures were identical for the two groups of women, all of whom were Tunisian.

Blood samples on tubes containing EDTA were collected from a venipuncture for extraction of genomic DNA after the women concerned had been informed of the nature of the study and its potential purpose, and their consent had been approved (**Appendix 2, Appendix 3**).

1.1.1 Patients

The patient population was made up of women under the age of 40 whose cause of ASR was not clinically clear.

The inclusion criteria are: two or more successive pregnancy losses in the first trimester from the same partner.

The exclusion criteria are: Rhesus incompatibility of the blood group, women aged over 40 in the first trimester of pregnancy, patients with endocrine disorders, diabetes, anatomical abnormalities and infections (Cytomegalovirus, Herpes simplex virus, measles, toxoplasma, syphilis, mycoplasma and bacterial vaginosis).

1.1.2 Witnesses

The female controls had had at least two children and had no family history of RSA. The controls were age-matched to the patients.

II. Methods

II.1 1 Blood sample

A blood sample on an EDTA tube had ële effects for each woman. Plasma sëparation a ëlë rëalisëe by centrifugation at 3000 revolutions (tr) for 15 minutes (min). The plasma obtained was discarded and the remaining blood pellet containing leucocytes was stored at -20°C for extraction of gënomic DNA.

II.2 2 Biochemical investigation

II.2.1 1 Determination of LDL cholesterol

The level of LDL cholestërol (LDLc) was calculated using the Friedwald formula based on blood levels of total cholestërol (TC), HDLc and triglycërides (TG):

LDLc (mmol/L) = CT (mmol/L) - (TG (mmol/L) /2,2) + HDLc (mmol/L))

The assay was performed on the Unicel® Dxc 600 (BeckmanCoulter).

II.2.2 2 Measuring blood glucose levels

The determination of glycëmia a ël ë rëalisëe according to a colorimëtric method with glucose oxidase on the Unicel® Dxc 600 automat (Beckman Coulter) :

Glucose oxidase

$$\text{Glucose} + H_2O + O_2 \xrightarrow{\text{Glucose oxydase}} \text{Acide gluconique} + H_2O_2$$

$$H_2O_2 + \text{phénol} + \text{4-AP} \xrightarrow{\text{Peroxydase}} \text{Quinone imine (rose)} + 4\ H_2O$$

4-AP: Amino 4-antipyrine

II.2.3 3 Protein assay

Protein assay is performed by Unicel® Dxc 600 (Beckman Coulter), using the Biuret colorimëtric method performed as follows:

Protéines + Cu^{2+} → complexe bleu violet

Proteins + Cu^{2+} ^ violet-blue complex

Cu^{2+} : Copper II ions

II.2.4 4 Dosing of 1'uree

The urea a ël.ë dosë using Unicel® Dxc 600 (Beckman Coulter), by an enzymatic method involving two hydrolysis reactions:

NH3: Ammonia

$$\text{Urée} + H_2O \xrightarrow{\text{Uréase}} 2NH_3 + CO_2$$

$$NH_3 + \alpha\text{-Ketoglutarate} + NADH + H^+ \xrightarrow[\text{GLDH}]{} \text{Glutamate} + NAD^+ + H_2O$$

NADH: Nicotinamide adenine dinucleotide reductase

GLDH: Glutamate dehydrogenase

II.3 3 Genomic DNA extraction

Genomic DNA is extracted from whole blood using the '*salting out*' method, which is based on a chemical and thermal method followed by purification with an enzyme, proteinase K, and a ëthanol precipitation step (Miller et al., 1988).

II.3.1 1 Lysis of red blood cells

Lysis of the red blood cells is carried out using SLR solution (Tris 1M, EDTA 0.5M) by hypotonic shock (**Appendix 1**). After centrifugation at 3000 rpm for 12 minutes, only the pellet is recovered.

Three washes by SLR are recommendedës in order to obtain a white pellet dëpourvu of any trace cП1ётод1оЬ1пег

II.3.2 2 Lysis of white blood cells

The pellet obtained is treated with SLB solution containing 10% dëtergent SDS (*Sodium Dodecyl Sulfate)* (**Appendix 1**), whose role is to destroy the cell membrane and release the cytoplasmic and nuclear contents. Nucleic DNA is released by treatment with proteinase K, which digests the histones associated with eukaryotic chromosomes. After homogenisation, the mixture is incubated for two hours at 55°C.

II.3.3 3 Protein precipitation and DNA purification

A volume of 666p.l of 6M NaCl (**Appendix 1**) is added to the resulting mixture of proteins and nucleic acids, followed by rigorous agitation and 10 min cold incubation. After

centrifugation for 15 min at 3000 rpm, the proteins are dehydrated and precipitated in the pellet.
2 to 3 volumes of cold absolute ethanol (100%) are added to the recovered supernatant to precipitate the DNA medusa. This is transferred to another 1.5ml tube and dried on the bench. Finally, the medusa was redissolved in 200pl of distilled water or Tris-EDTA solution to ensure better preservation of the extracted DNA.

II.3.4 4 DNA purity control and quantification

Once the DNA has been extracted, quality control is essential to ensure reliable amplification in subsequent analyses. In addition, precise knowledge of the concentration of extracted DNA is essential in order to make the necessary dilutions. This quality control was carried out using QIAGEN's Nanodrop QIAxpert instrument. The Nanodrop QIAxpert generates an absorbance curve, evaluates the concentration of the DNA sample analysed and provides relevant reports on the purity of the DNA by detecting possible contamination by proteins or residual traces of solute. DNA has a maximum absorbance peak at 260 nm, while protein absorbance peaks at 280 nm. The ratio R= A260/A280 is a good indicator of DNA purity with a target range between 1.6 and 2 for DNA to qualify as pure. A value below 1.6 indicates possible protein contamination, while a value above 2 indicates RNA contamination.

II.4 4 DNA sequence research and primer development

The sëquences containing the polymorphisms ëtudiës, as well as the appropriëed enzymatic sëquences, have ël ë searchedëes on the sites of the *"National Center for Biotechnology Information"* (www.ncbi.nlm.gov) and *New England Biolabs* (ww.neb.com). Knowing the compete nucleotide sëquences of our gënes of interest as well as the location of the corresponding polymorphisms, we were able to proceed with the design and Гadëquation of primers for amplification by the polymërisation chain reaction (PCR) of the gënëtiques ëtudiëes variants using the bioinformatics tools and database provided by the American University of *Santa-Cruz.*

II.4.1 SNP localisation and primer design

Step 1: Search for the gene sequence containing the polymorphic site

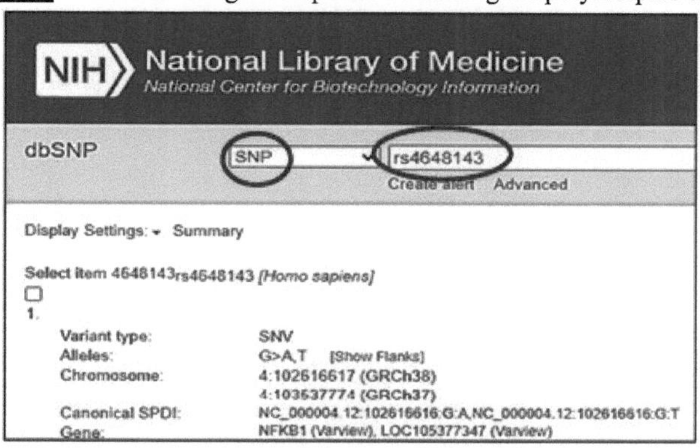

Figure 15: NCBI search for the genomic location of a polymorphism

Step 2: Copy the nucleotide sëquences located on either side of the SNP

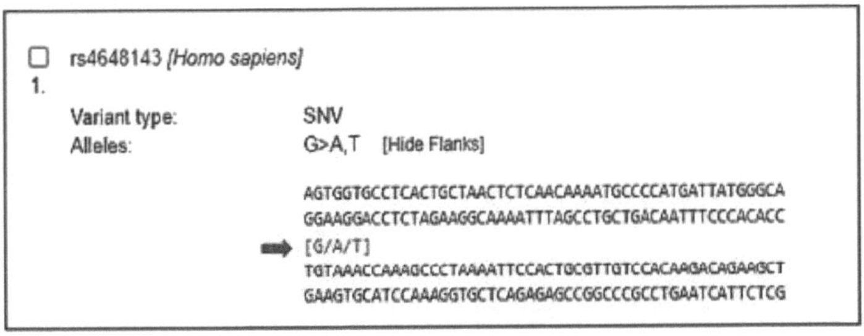

Figure 16 Dĕtermination of the gĕnĕtic variation of a SNP

Step 3: Determination of primer pairs by NCBI Primer-Blast

The selection of primers was made according to well-established criteria:

S A size between 20 and 30 bp

S GC composition > 50%

S Melting temperature (Tm) >55°C

S Difference between Tm of sense and antisense primers < 5°C

S Sequences perfectly complementary to the fragment to be amplified (absence of mismatches)

S Absence of self- or inter-coupling between primers, particularly in the 3'.

Step 4: Determination of the restriction enzyme using Nebcutter from *New England Biolabs*

This database can be used to identify the different restriction sites on the given nucleotide sequence for several enzymes. This makes it possible to determine the appropriate enzyme system (creation or abolition of the restriction site) by comparing the restriction maps of the ancestral and minor alie.

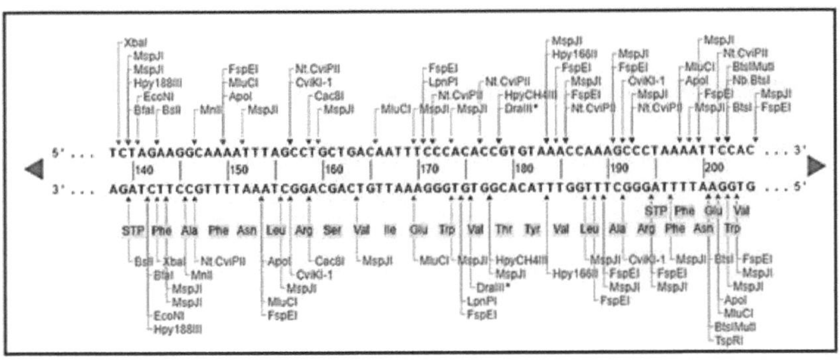

Figure 17: Identification of restriction enzymes

II.4.2 Genotypic analysis of genetic markers

We have selected the gĕnique variants at the *NFκBI* gene level suspected of susceptibility to 1'ASR based on the results of previous ĕ studies. These markers are a SNP-like point variation and are located in the 3'UTR region of the gene.

PCR-RFLP was used to study these SNPs.

11.4.2.1 SNP rs4648143

This variant is a G/A transition in the *NFkB1* gene on chromosome 4 at position 102616617. A 198 bp sequence was amplified by PCR using the following primer pair: **Forward primer:**

5'- CGCAAACTCAGCTTTACCGA-3'
Antisense primer : 5'- ACCTTTGGATGCACTTCAGC-3'
This transition leads to the appearance of a restriction site for the **DraIII** enzyme.

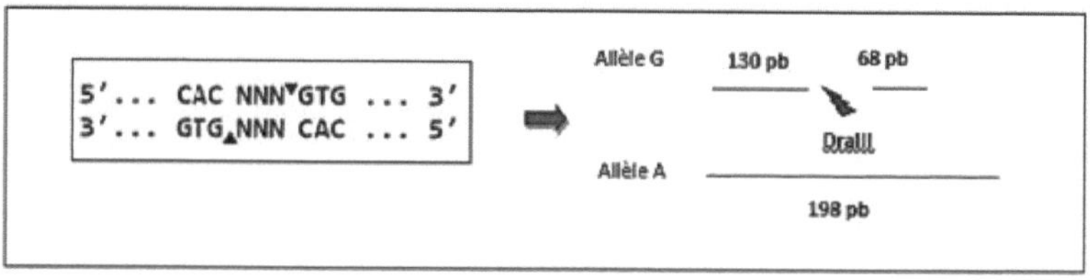

Figure 18: Restriction site and mapping of the rs4648143 gene variant using the DraIII restriction enzyme.

Description: The restriction enzyme DraIII cuts, at cohësive ends, at the ancestral G allele giving rise to two DNA fragments of size 130 bp and 68 bp respectively. When the relevant sëquence, also called amplicon and 198 bp in size, contains the mutated al^le (A), the enzyme is no longer able to cut since this al^le causes abolition of the DraIII restriction site. Thus, after enzymatic digestion, we obtain a single band corresponding to a fragment of the same size as the amplicon.

11.4.2.2 SNP rs41275743

This polymorphism consists of a G>A transition located on chromosome 4 at position 102616698. A 209 bp sequence was amplified by PCR using the following primer pair:
Primer sense : 5'- ACACCGTGTAAACCAAAGCC -3'
Antisense primer : 5'- CAGCCAGTGTTGTGATTGCT-3'
This transition leads to the abolition of the restriction site of the **HaeIII** enzyme.

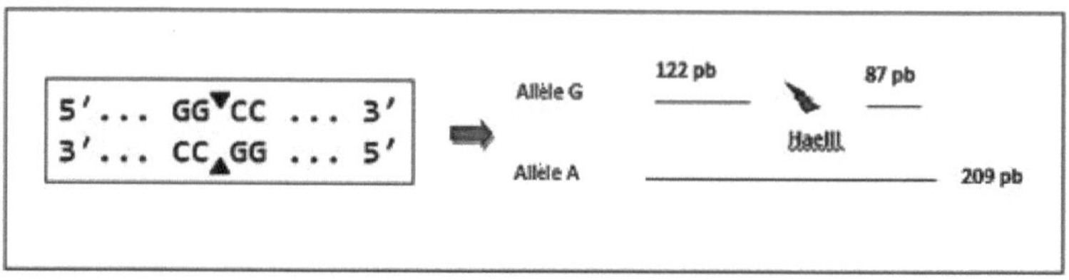

Figure 19: Restriction site and mapping of the rs41275743 gene variant by the 1 HaeIII restriction

Description: This enzyme cuts in the presence of the ancestral G allele, which contains its restriction site, thus producing two open-ended fragments of size 122bp and 87bp respectively.

11.4.2.3 SNP rs28362491 INDEL

This is an ATTG insertion/deletion polymorphism located in the promoter region of the *NFκBI* gene (-94 ins/del ATTG) on chromosome 4. A 285 bp amplified sëquence from the following primer pairs:
Forward primer: 5'-TGGGCACAAGTCGTTTATGA-3'
Antisense primer: 5'-CTGGAGCCGGTAGGGAAG-3'
Insertion results in the creation of a restriction site for the **PflMI** enzyme.

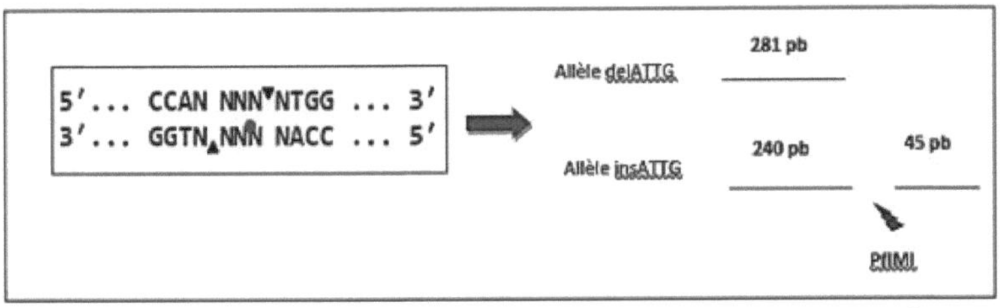

Figure 20: Restriction site and mapping of the gënic variant rs28362491 INDEL by **PflMI** restriction enzyme

Description : The PflMI enzyme cuts in the presence of an ATTG insertion which creates a restriction site for this enzyme, resulting in two fragments with cohesive ends of 240 bp and 45 bp respectively.

11.5 Reaction medium and PCR

PCR amplification reactions are one of the greatest discoveries in molecular biology. It is a method that makes it possible to exponentially increase the number of copies of genetic material existing in a biological sample, even in very small quantities. It consists of a succession of thermal cycles during which the number of DNA molecules, acting as a matrix, is doubled with each cycle.

After "n" cycles of PCR, we theoretically have "2^n " copies of identical, amplified DNA called amplicons. The PCR reaction takes place in a thermocycler for 30 to 40 cycles, during which the temperature varies according to the amplification stages: denaturation (95°C), hybridisation (50-65°C) and elongation (72°C), which take place 35 times.

Each PCR tube contains :

- Buffer : 10X
- MgCl2 : 20mM
- DNTP : 25 mM
- Amorce sens : 10 µM/ µl
- Amorce antisens : 10 µM/ µl
- Taq DNA polymérase : 5U
- ADN : 100 ng

5 min à 95°C
40s à 94°C
40s à Ta*
40s à 72°C } X35
7min à 72°C

Ta*: (*annealing temperature*) or hybridization temperature (Th).

Ta of rs4648143 = 59°C / Ta of rs41275743 = 59°C / Ta of rs696 = 62°C / Ta of rs28362491 = 57°C.

11.6 Verification of amplification

This ëstep ensures the correct dëroulement of the DNA amplification before moving on to enzymatic digestion by vërifying the presence of the PCR products of our ëcЬaпйПoпз by ëlectrophorëtic migration, by subjecting them to a ëelectric field on a 2% agarose gel (**Appendix 1**). The iregatively charged nucleic acids move through the gel matrix from the cathode (-) to the anode (+).

The amplified DNA fragments are easily visualised on the gel using Ethidium Bromide (BET), which is an intercalating component capable of binding between the nucleic bases inside the double helix and a dye that emits red-orange fluorescence following excitation by ultraviolet light.

11.7 Genotyping and enzymatic digestion

Using restriction enzymes, the gënome of an individual will be cc^ë into several fragments depending on the number of restriction sites of the enzyme usedë. The number of these sites and their positions diffërent from one individual to another. The PCR products will be digërësed by restriction enzymes spëcific to each polymorphism and the fragments thus obtained will be sëparësed according to their sizes by migration on 2% agarose gel, allowing the dëtermination of the diffërents gënotypes corresponding to each SNP.

Each tube contains the following ëlëments: 1.5 Щ of buffer (10X), 0.18 |il of enzyme,

5.32 Щ of water, and finally, 5 Щ of the PCR product containing the target DNA. This reaction mixture is then incubated at 37°C, creating an environment conducive to specific DNA amplification.

II.8 Statistical analysis

Data entry and statistical analysis were carried out using the following three biostatistical software packages:

1. SPSS version 24.0 (SPSS Inc., Chicago, IL, USA)
2. SNPstats (https://www.snpstats.net/)
3. Haploview 4.2 (http://www.broad.mit.edu/mpg/haploview)

After having ëtudië and compared the distribution of modal^s within the groups of tëmoins and patients, as well as according to their own nature, the various clinical and biological variables will be examined. This will be done either in dichotomous form (normal/abnormal, present/absent), or in continuous form, or after catëgorisation according to their membership of several distribution intervals dëfmis on the tëmoin group. The modality catëgorisation will be performed for each continuous variable by comparing the distribution of percentiles between the control and patient groups. Indeed, a divergence between these two groups indicates the threshold at which it becomes relevant to characterise groups of values that could potentially present a risk. The differences are statistically significant when *the P* value is less than 0.05.

In the general population, the genotypic distribution of a SNP must conform to the Hardy-Weinberg equilibrium principle.

11.8.1 Hardy-Weinberg equilibrium

Hardy-Weinberg equilibrium (HWE) was assessed for each SNP using a statistical livpotliese test based on the distribution of a % variable^2 (Chi-carre test). A *P* value > 0.05 means that the observed and expected genotype distribution values are not significantly different and that the polymorphism under study is in HW equilibrium. The associated *P values* are obtained from the SNPstat software.

11.8.2 Haplotype analysis

A haplotype is a combination of alleles from different loci located on the same chromosome and generally transmitted together. Theoretically, for n biallelic SNPs there are 2^n possible haplotypes. Today, it is possible to estimate haplotype frequencies within a study population, based on measured genotype frequencies, thanks to bioinformatics software such as Haploview, which makes it possible to assess the impact of a particular combination of polymorphisms on the risk of developing the disease, rather than focusing on an individual polymorphism. In the event of linkage disequilibrium, nucleotide sequences are transmitted in the form of haplotypic blocks, thereby considerably reducing the probability of recombination within a single block. Haplotypic analysis therefore offers the possibility of assessing the effect of a particular combination of polymorphisms on the risk of developing the disease by using a limited number of SNPs in genetic studies (Barrett et al., 2004; Montpetit and Chagnon, 2006).

II.9 Epigenetic study

11.9.1 Sampling

Participants in this study were ële sëlectedë from women with a history of 4 or more spontaneous abortions. A 42-year-old female patient with a history of 7 recurrent spontaneous abortions with no identified cause was recruited. A control woman of the same age, with at least two children and no history of ASR, was also included. All were of Tunisian origin. Blood samples were taken by venipuncture on EDTA tubes in order to extract total RNA. The women concerned approved their consent to participate in our study after having been informed of these potential objectives (**Appendix 3**).

11.9.2 Extraction of total RNA by TRIzol

In order to study the expression of hsa-miR-21-5p and a group of its target genes involved in the inflammation pathway in the two groups of women, total RNA was extracted using a method based on the differences in physico-chemical properties between the different nucleic acids (DNA and RNA) and proteins. The reference method applied for this extraction is based mainly on the use of a chemical called TRIzol, which is a single-phase solution composed of guanidinium isothiocyanate, a powerful protein denaturant, and phenol, an organic solvent.

Red blood cells from the blood sample that have not passed the three-hour time limit are eliminated using a hypotonic lysis solution (SLR). Incubation for 10 min at -20°C and centrifugation at 3500 rpm for 7 min at +4°C allows precipitation of the white blood cells. After carrying out two repetitions of this washing step, each time followed by elimination of the supernatant, TRIzol was added to the pellet followed by manual shaking and then transferred to another 1.5 ml tube to which chloroform was added. The mixture was shaken vigorously by vortexing, incubated for 10 min at -20°C and centrifuged for 15 min at 12000 rpm at +4°C.

The homogenate is thus separated into three phases, an organic phase at the bottom of the tube containing the proteins, an interphase layer containing the DNA and an upper transparent aqueous phase containing the RNA which will be precipitated by isopropanol. Sufficient 75% ethanol is added to precipitate the DNA and the proteins are precipitated from the phenolethanol supernatant after centrifugation for 10 min at 7000 rpm. The RNA pellet is sëchë briëvement in open air and solubilisë by addition of ultrapure water.

In order to remove any DNA present in the mixture, DNasel (10gl) is used and the mixture is incubated at 37°C for 10 min. The action of DNasel is stopped by the addition of 7ɥl EDTA 10X (50mM; pH8) followed by incubation for 10 min at 75°C.

Quantification by optical density measurement is performed to ensure the purity of the RNA obtained, which can then be stored at -80°C.

11.9.3 RT-qPCR

qPCR, also known as quantitative PCR or real-time PCR, is a technique derived from conventional PCR for measuring the initial quantity of DNA. RT-qPCR is based on the

amplification of an extracted RNA sequence used as a starting template after it has been transcribed into complementary DNA (cDNA) by a reverse transcription process to form a DNA-RNA hybrid. The RNA strand is then degraded, leaving only the cDNA strand, which is used as a template for conventional PCR amplification.
RT-qPCR is performed using either SYBR Green or TaqMan, in two stages involving two separate reactions:

11.9.3.1 Reverse transcription

11.9.3.1.1 By M-MLV

Reverse transcription is performed by Moloney murine leukemia virus (M-MLV) reverse transcriptase (RT), which is an enzyme capable of synthesising cDNA. In a microcentrifuge tube, different components are added according to the protocol: 1 Щ of Oligo(dT) (500^g/ml) is incorporated, followed by the addition of 1 Щ of dNTP (10mM). Next, 2 Щ of RNA is introduced. Equilibrium of the mixture is reached by adding 8 Щ of water. This mixture is incubated at 65°C for 5 min and then rapidly cooled on ice to stop the reaction. The contents of the tube are then recovered by a breve centrifugation to complete the preparation of the reaction mixture by adding 1 Щ of 5X First-Strand Buffer, 2 Щ of DTT (100mM), and 1 pl of RNase OUT™ (40 units/pl.), thus guaranteeing protection against enzymatic dëgradation of the RNA.
After gentle shaking of the tube, the mëlange is incubated at 37°C for 2 min. Next, 1 pl of M-MLV (200 units (U)) is added and thoroughly mixed by gentle pipetting. Incubation at 37°C for 50 min and then at 70°C for 15 min is performed to inactivate the enzyme and the cDNA is finally synthesised, which will be used as a template for PCR amplification.

11.9.3.1.2 TaqMan™ Small RNA Assays

Applied Biosystems™ TaqMan™ reagent kits are designed to detect and quantify a variety of small RNAs such as mature microRNAs, interferent RNAs, and others. When used to analyse microRNAs, these kits can distinguish the sequences of mature microRNAs from those of their precursors. Each TaqMan™ Small RNA Assays kit comprises two tubes: one containing a stem-loop RT primer for reverse transcription, and the other containing a mixture of RNA-specific sense and antisense PCR primers and a specific TaqMan™ probe.
Each tube should contain a 7pl volume of reverse transcription reaction mix containing: 100mM dNTP, MultiScribe™ Reverse Transcriptase (50 U/pL) ,10X Reverse Transcription Buffer, RNase Inhibitor (20 U/pL) and nuclease-free water. A volume of 5 pl of RNA is then added, and these 5 pl should contain between 1 and 10 ng of RNA. After homogenisation and centrifugation, 3 pl of RT primer is added to the mixture. After centrifugation and placing the tubes on ice, the reverse transcription reaction is carried out in a thermocycler, first at 16°C for 30 minutes, then at 42°C for the same length of time, and finally at 85°C for 5 minutes in order to stop the action of the transcriptase.

11.9.3.2 The qPCR reaction

11.9.3.2.1 By SYBR Green

SYBR Green is a fluorescent dye with the ability to bind to nucleic acids and intercalate between double strands. It is used to quantify the quantity of amplicons during PCR by emitting fluorescence. SYBR Green is frequently used for non-spëcific detection. SYBR Green I is a molecule that can bind to different types of double-stranded nucleic acids. When Hë, it remarkably increases its fluorescence. Thus, the db DNA/ SYBR Green I complex absorbs blue lumiëre with a maximum wavelength (Imax) ёдак at 497 nm and ëmet green lumkre at Imax = 520 nm. The fluorescence ë emitted is proportional to the quantik of

amplicons.
An ROX rëfërence dye is an inert additive that provides fluorescent signal normalisation and correction of optical variations in each well throughout qPCR.

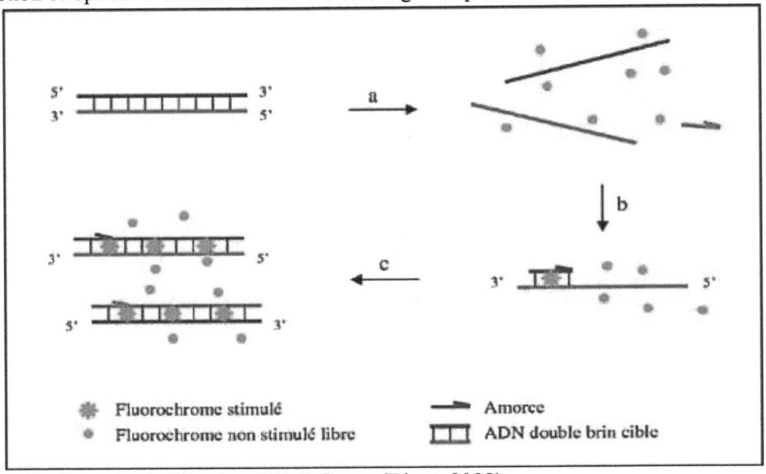

Figure 21: Steps in qPCR using SYBR Green (Bigot, 2009)

Description: During dënaturation (a), SYBR Green is free and unable to bind to single-stranded DNA, so it exhibits weak fluorescence. During hybridization (b), the molecule intercalates between nascent DNA double strands resulting in an increased ëmission of fluorescence after excitation. During elongation (c), an increasing number of SYBR Green bind to DNA amplicons db thus Смшё by ëmitting strong fluorescence dëtectëe in real time.

11.9.3.2.2 By hydrolysis of probes: Taqmanassay

TaqMan technology is based on the 5'-exonuclease activik of the Taq polymërase enzyme, which allows the hydrolysis of a fluorogenic probe hybridised specifically to the target sequence on the amplicon during hybridisation and PCR extension. This technology is used to detect only specific amplification products during PCR cycles.

Its principle is basë on the construction of an oligonucteotide probe containing a ëтeйeиг fluorochrome, appelë Reporter, йxë at the 5' end of the hybridization probe. When stimulated, its ëmission is inhil^ed by another suppressor fluorochrome called Quencher, sitiro at the 3' extrëmitë of the probe.

The amplification cycle comprises two phases:

- **The hybridization phase:** during which the probe and primers bind to their target sequences by basic comptementarE
- **The elongation phase**: synthesis of a new strand of cDNA from the primer

by Taq polymërase. When the latter encounters the hybridëed probe during polymërisation, it dëplaces it and hydrolyses it using its 5' exonuctease activity, the Reporter is then Hьërë thus enabling Emission of fluorescence which increases with each cycle in proportion to the rate of hydrolysis of the probe. Thus, Elongation continues until the end of the target sëquence.

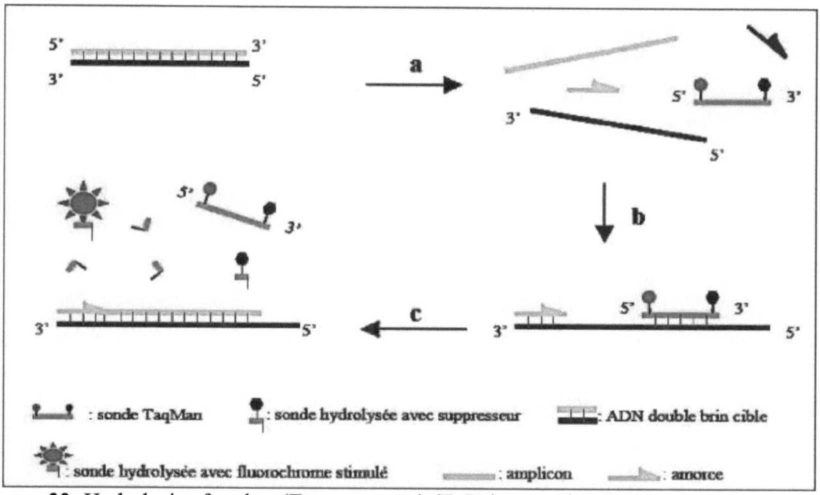

Figure 22: Hydrolysis of probes (Taqman assay) (E. Poitras et al., 2002)

<u>Description</u>: During dënaturation (a), the TaqMan probe (in purple) is free in solution. At hybridization tempërature (b), the probe and primers (in yellow) hybridize to their respective target sëquences and the proximity of the fluorochromes (the small circles in black and red above the probe) allows fluorescence to be inhibited. When polymerisation begins, Taq polymerase displaces and hydrolyses the probe (c) and the emitting fluorochrome (reporter, the small red circle) is released from its inhibitor (quencher) and thus emits fluorescence detected by the qPCR apparatus.

11.9.3.2.3 Relative quantification

Relative quantification (Rq) is a frequently used principle in gene expression studies and is based on the comparison of threshold cycles (Ct), which measures the difference in expression of the target gene in the sample compared to a reference gene. Firstly, the difference between the Ct values (ACt) of the gene of interest and the reference gene is calculated for each experimental sample. Next, the difference in ACt values between the patient and control samples (AACt) is calculated.

The *Fold change* (Fc) or Rq of the expression of a gene of interest relative to the control is equal to 2^{-AACt} . A result is considered significant when the Fc is at least two, equivalent to an Rq greater than 2 or less than 0.5.

11.9.3.2.4 qPCR reaction medium: SYBR Green

After gentle shaking and brief centrifugation of the thawed solutions, each tube should contain a total volume of 25 gl. The reaction mixture is prepared by adding the following reagents:

- Maxima SYBR Green qPCR Master Mix (2X), with ROX: 12.5 gl
- Sensory primer: 0.3 gM
- Antisense primer: 0.3 gM
- Nuclease-free water: a 25 gl

After homogenisation, the appropriate volumes are divided into different tubes, and the corresponding cDNA, the quantity of which is maintained between 50 and 500 ng, is added to each tube. The instrument's thermocycler is programmed to run 40 cycles according to the following protocol: retrotranscription takes place at 95°C for 10 min. Then, during the amplification phase, the temperature is maintained at 95°C for 15 seconds, followed by an elongation step at 60°C for 1 minute. Finally, the melting curve is generated in the last step, with temperatures of 95°C for 15 seconds and 60°C for 1 minute. These parameters are carefully adjusted to ensure an efficient reaction throughout the process.

II.9.3.2.5 Reaction medium by TaqMan

Each well of the reaction plate is designed to hold a total volume of 20 gl, incorporating the following components: 1 Щ of TaqMan™ Small RNA Assay (20X), 10 Щ of TaqMan™ Universal PCR Master Mix, without AmpErase™ UNG (2X) which is included to create optimal reaction conditions. The cDNA, which is strictly limited to 1.33 gl, represents the maximum volume that can be added to the qPCR reaction. It is crucial that the amount of cDNA remains within the range of 1 to 100 ng. Finally, 7.67 gl of water is added to reach a total volume of 20 gl. Once the preparation is complete, the plate is briefly centrifuged to remove air bubbles from the solutions and loaded into the qPCR system according to a programme that includes the activation of Uracile N-glycosylase (UNG), an optional step carried out at 50°C for 2 minutes, allowing the removal of uracil from the DNA. The DNA polymerase is then activated at 95°C for 10 min. The reaction continued with denaturation at 95°C for 15 seconds, followed by hybridisation and elongation at 60°C for 60 seconds.

CHAPTER 3

Results

I. Clinical and biochemical characteristics of the population studied

Our study includes a group of 243 patients who had at least two consëcutifs and unexplained abortions and a group of 251 control women who had at least one normal pregnancy without antdcddents of abortion. The clinical and biochemical characteristics are shown in **Table I**. We found a statistically significant difference in mean age (P<0.001), BMI (*P=0.*001), number of live children (*P<0.*001), number of abortions (*P<0.*001), caffeine intake (*P=0.*046) and similarly for LDL *(P=0.*010) and protein (*P=0.*001) levels. However, patients and controls were matched on oral contraceptive use (*P=0.*144), blood glucose *(P=* 0.648) and urea (*P=0.*355).

Table I: Clinical and biochemical characteristics of patients and controls

	Patients	Witnesses	Value of P[4]
Age (years) [1]	31.90 ± 5.56	33.89 ± 6.29	**< 0.001**
Body mass index (kg/m2) [1]	25.3 ± 4.1	26.6 ± 5.62	**0.001**
Number of living children IQR (min-max) [2]	1 (0-5)	2 (0-7)	**< 0.001**
Abortions IQR (min-max) [2]	2 (2-12)	0 (0-1)	**< 0.001**
Use of oral contraceptives (n [%]) [3]	10 (3.6)	24 (6.2)	0.144
Caffeine consumption (n [%]) [3]	125 (52.7)	188 (61.8)	**0.046**
Smoking (n [%]) [3]	53 (16.7)	98 (25.1)	**0.006**
Glucose (mmol/l) [1]	5.05 ± 0.79	5.14 ± 1.62	0.648
LDL (mmol/l) [1]	2.26 ± 0.84	2.66 ± 1	**0.010**
Protein (g/l) [1]	68.2 ± 6.58	71.8 ± 6.36	**0.001**

I 1: Mean ± standard deviation, **2:** Median IQR (minimum-maximum), **3:** Number of subjects as a percentage and **4:** Student's t test for continuous variables, Pearson's Chi-square test for categorical variables.

II. Allelic and genotypic distribution of *NFκBI* gene polymorphisms

II.1 Allelic frequencies

The rs41275743 ëtudië polymorphism of the дпе *NFkBI* ële conformably distributed at Hardy Weinberg equilibrium in both the tëmoin group and in patients presenting with ASR as 1 indicated in **Table II** (HWE=0.35). This ëqшlibre has ël.ë testë for all three SNPs by SNPstats software. We also calculated the power for the SNPs in question using *QUANTO* 1.2 bioinformatics software. In our study population, the altered frequency distribution of polymorphisms rs41275743, rs4648143 and rs28362491 showed a significant diffërence between patients and tëminors; rs41275743 (*P* = 0.013; OR = 0.52 (0.29-0.95)), rs4648143 (*P* <0.001; OR = 0.36 (0.27-0.47)), rs28362491 (*P* <0.001; OR = 1.84 (1.39-2.45)).

Table II: Distribution of *NFκBI* gene alleles

SNP	Position	Percentage of	Power (%)	Allele	HWE	Patients' FAM	FAM des	P[1]	OR (95%

		genotyping (%)					temoins		CI)
rs41275743	102616698	93.72	80%	G/A	0.35	17 (0.04)	35 (0.07)	**0.013**	0.52 (0.29-0.95)
rs4648143	102616617	92.11	80%	G/A	0.00076	106 (0.25)	230 (0.48)	**<0.0001**	0.36 (0.27-0.47)
rs28362491 INDEL	-94	95.55	80%	del ATTG /dup ATTG	<0.0001	158 (0.36)	115 (0.23)	**<0.0001**	1.84 (1.39-2.45)

MAF: Minor allele frequency, **OR:** Odds ratio, **CI:** Confidence interval, ***P***: *P* value, **1.** Pearson Chi-square test

II.2 Genotype frequencies

The gënotypic distributions of rs41275743, rs4648143 and rs28362491 are compared between the two study groups as shown in **Table III** and according to three models of gënëtic inheritance (codominant, dominant and recessive) in **Table IV**, the results of which confirm the association of rs41275743, rs4648143 and rs28362491 with RSA.

11.2.1 Genotypic distribution of the rs41275743 polymorphism

The genotypic distribution shows the presence of an association of the rs41275743 polymorphism and the risk of ASR (*P=0*.013) with a gënotypic frequency of 5% of the hëtërozygote G/A gënotype and 1.4% of the minor A/A gënotype in patients. In tëminors, a frequency of 12.7% G/A gënotype and 0.8% A/A gënotype. Both the codominant and dominant models were associated with ASR, unlike the recessive model, which showed no association with ASR risk.

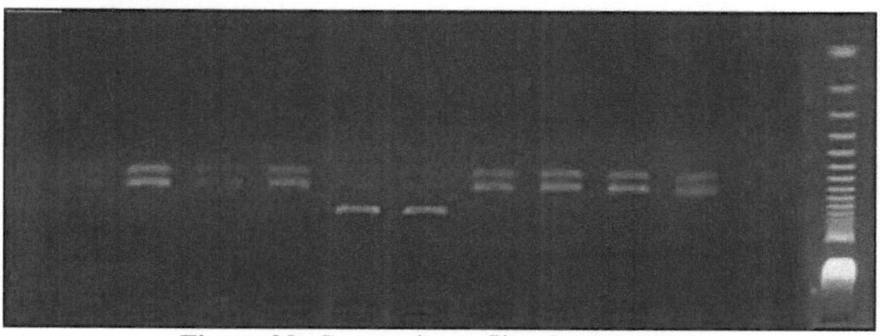

Figure 23: Genotypic profile of rs41275743

11.2.2 Genotypic distribution of the rs4648143 polymorphism

The genotype distribution showed an association between the SNP rs41275743 and the risk of ASR (*P<0*.001) with a genotype frequency of 41.1% of heterozygous genotype G/A and 4.2% of minor genotype A/A in patients. In the controls, a frequency of 39% of the G/A genotype and 28.2% of the A/A genotype. The three codominant, dominant and recessive models showed a significant association with the disease.

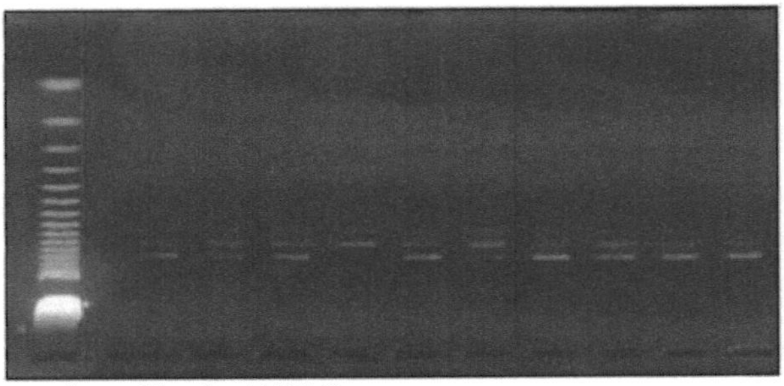

Figure 24: Genotypic profile of rs4648143

11.2.3 Genotypic distribution of the rs28362491 INDEL polymorphism

The genotype distribution showed an association between rs28362491 and the risk of ASR ($P<0.001$), with a genotype frequency of 57.7% for the heterozygous del/ins genotype and 6.8% for the minor ins/ins genotype in patients. In controls, a frequency of 44.4% of the del/ins genotype and 0.8% of the ins/ins genotype. The three codominant, dominant and recessive models are associated with ASR.

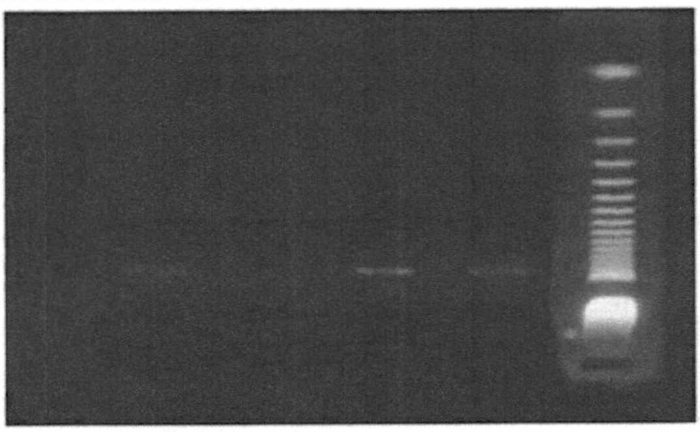

Figure 25: Genotypic profile of rs28362491

Table III: Distribution of *NFκBI* gene gënotypes

	1/1 [4]		*1/2* [4]		2/24			
SNP	**Patients**	**Witnesses**	**Patients**	**Witnesses**	**Patients**	**Witnesses**	***P*** [II]	***aP*** [2]
rs41275743 (G/A)	204 (93.6)[3]	212 (86.5)	11 (5)	31(12.7)	3 (1.4)	2 (0.8)	**0.013**	**0.041**
rs4648143 (G/A)	117 (54.7)	79 (32.8)	88(41.1)	94 (39)	9 (4.2)	68 (28.2)	**<0.0001**	**<0.0001**
rs28362491 (del/ins)	79 (35.6)	137 (54.8)	128 (57.7)	111 (44.4)	15 (6.8)	2 (0.8)	**<0.0001**	3e-04

Table IV: Gënotypic distribution of SNPs ëtudiës according to gënëtic modëles.

II Pearson Chi-squared test, ***aP2***: P adjusted for age and BMI and **3.** Gënotypic frequency in percentage and **4.** Gënotype dësignës by *1* "wild allele" and *2* "mute allele".

Codominant model				Dominant model		Recessive model	
SNP	Genotype	*Value of* *1P*	OR (95% CI)	*Value of* *1P*	OR (95% CI)	*Value of* *1P*	OR (95% CI)
rs41275743	G/G	0.013	1.00 (Rëfërence)	G/G vs. G/A + A/A		G/G + G/A vs. A/A	
	G/A		0.37 (0.18-0.75)	0.011	0.44 (0.23-0.85)	0.56	1.70 (0.28-10.24)
	A/A		1.56 (0.26-9.43)				
rs4648143	G/G	<0.0001	1.00 (Rëfërence)	G/G vs. G/A + A/A		G/G + G/A vs. A/A	
	G/A		0.63 (0.42-0.95)	<0.0001	0.40 (0.28-0.59)	<0.0001	0.11 (0.05-0.23)
	A/A		0.09 (0.04-0.19)				
rs28362491	del/del	<0.0001	1.00 (Rëfërence)	del/del vs. ins/del+ ins/ins		del/del + ins/del vs. ins/ins	
	ins/del		2.00 (1.37-2.91)	<0.0001	2.19 (1.51-3.18)	3 x10-4	8.9 (2.03-39.75)
	ins/ins		13.01 (2.90-58.36)				

1. Pearson Chi-squared test

III. Haplotype analysis

Haplotypes of the *NFκBI* gene are constructed based on the interaction of the three SNPs (rs28362491, rs4648143, rs41275743) and linkage disequilibrium analysis between them. Of the eight possible haplotypic combinations, only six haplotypes were ë1.ë present in our Tunisian population (**Table V**).

Haploview analysis demonstrated that the combination of the 3 ëtudiës polymorphisms, in a block covering 115 kb, showed a linkage disequilibrium (d') between these three variants and a significant association in the frequencies of the four haplotypes between patients and controls. For technical reasons of the Haploview software which considersëonly one allele, the deletion or insertion of the ATTG nucleotide of rs28362491 was replaced by the T allele (del= T) and the A allele (ins=A) respectively. A significant association of the TAG (*P=4.*5**10**e-11), AGG *(P=3.***8037** "e7) and TAA (*P=0.*0224) haplotypes has ërë observed.

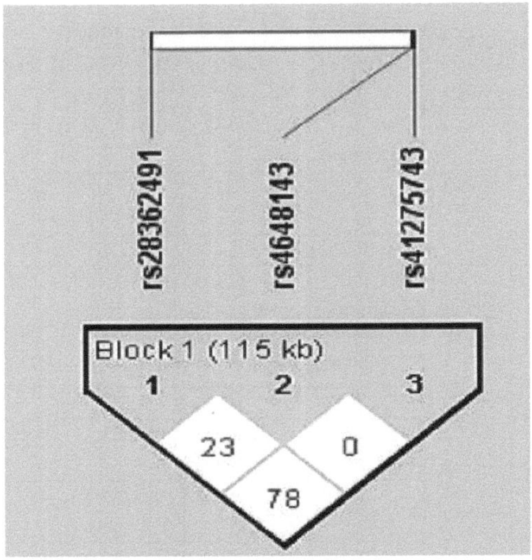

Figure 26: Map of dësëbinding equilibrium (LD) of polymorphisms in the *NFκB1* gene analysëe by Haploview.

Table V: The different haplotypic forms of the three SNPs of the *NFκB1* gene

Haplotypes	Total[1]	Patients[1]	Witnesses[1]	x^2	*P*
TGG[2]	0.389	0.475	0.342	8.172	- --
TAG	0.268	0.128	0.363	43.379	**4.5104**$^{e-11}$
AGG	0.209	0.242	0.142	25.791	**3.8037 7**$^{e-}$
AAG	0.079	0.114	0.080	0.048	0.8264
TGA	0.034	0.034	0.038	0.391	0.5315
TAA	0.019	0.005	0.026	5.212	**0.0224**

1 : Haplotypic frequency, **2 :** The most frequent haplotype in female controls (TGG) was taken as the reference (OR=1.00) and ***P* :** Haplotypic association, *P* value < 0.05.

IV. Analysis of qPCR results

I . Expression levels of *NFκBI* and *IL-6* genes were assessed in the patient and female control using SYBR Green, while the specific expression of hsa-miR- 21-5p relative to its target genes was assessed using TaqMan. These analyses were performed in duplicate for each subject. The results obtained are presented in detail **in Table VI.**

Table VI: Quantitative analysis of *NFκBI* and *IL-6* gene expression and of hsa-miR-21-5p

	Sick		**Timer**				**Level of**
	Average Ct	**A Ct**	**Ct**	**A Ct**	**AACt**	**Rq**	**expression**
NFκBI	24.733	2.031	23.512	0.832	1.199	0.435	Sub-expression
GAPDH [1]	22.702		22.680				
IL-6	21.834	-0.868	21.120	-1.56	0.692	0.618	Normal
GAPDH	22.702		22.680				
MiR-21-5p	22.51	-1.77	24.00	-0.322	-1.448	0.408	Sub-expression
RNU6B [2]	24.280		24.322				

1 .Gene reference for SYBR Green

2 .MiRNA reference for TaqMan

ACt Sick = Ct Sick - Ct Gene reference

ACt Temoin = Ct Temoin - Ct Gene de reference

AACt = Sick ACt - Control ACt

$$Rq = 2^{-AACt}$$

CHAPTER 4

Discussion

Spontaneous and unexplained pregnancy losses can be emotionally and physically devastating for couples, particularly when they are recurrent, and they present a frustrating reproductive problem for both the patient and the doctor.

Despite the great interest in RSA as one of the most common reproductive complications, the definition of RSA remains controversial. Several studies have suggested that the causes of miscarriage identifiable after two successive pregnancy losses are similar to those identified after three consecutive losses (Kutteh and Stephenson, 2006; Turesheva et al., 2023).

Thus, ESHRE prĕcĕdently defined RSA as three consĕcutive pregnancy losses before 20eme SA, whereas ASRM defined RSA as at least two pregnancy losses (Jauniaux et al., 2006; ASRM, 2012). According to the ESHRE's latest update of this meaning in 2017 and 2022, it considers, like the ASRM, that RSA is the occurrence of two or more pregnancy losses (ESHRE, 2017; Bender Atik et al., 2023).

Genetic variations appear to play a complex role in human reproductive efficiency. Recent research has highlighted the crucial role of genetic polymorphisms in maintaining pregnancy, paving the way for ĕstudies of phĕnotypic and gĕnotypic associations to better understand the pathophysiological mĕcanism behind RSA (Hyde and Schust, 2015).

Inflammation is a natural response of the immune system to an aggression of the organism, inflammatory processes may play a role in idiopathic RSA considering their involvement in feto-maternal tolerance and good pregnancy outcome (Comba et al., 2015).

NF-κB is a transcription factor that plays a central role in the inflammatory response. It ensures proper placental development by regulating the activity of pro- and anti-inflammatory cytokines expressed during the different trimesters of pregnancy (Armistead et al., 2020; Gomez-Chavez et al., 2021). Excessive activation of NF-κB leads to an immune rejection response in the fetus. Consequently, NF-κB could be a potential target for the treatment of recurrent abortion (Zhang et al., 2021).

Evidence for genetic contributions to obstetric disorders comes mainly from candidate gene-based association studies, where researchers have evaluated particular polymorphisms of genes sĕlectnĕs mainly because of their potential role in pregnancy such as the *NFκBI* gene. Most of the polymorphisms studied are considered functional since they influence phenotypic expression (Daher et al., 2012).

Several studies have focused on the alteration of дёпе *NFκBI* expression in relation to spontaneous abortion (Gomez-Chavez et al., 2021; Logan et al., 2022).

In the present study, we focused on the evaluation of three polymorphisms of the *NFkB1* gene, rs41275743, rs4648143 and rs28362491, in patients with ASR and multiparous women. The polymorphisms in our study have not previously been investigated in the context of RSA, which reinforces the importance of our research.

The SNPs rs41275743 and rs4648143 are located in the 3' UTR region of the *NFκBI* gene, which could influence the affinity of certain miRNAs and modify *NFkB1* expression (Zhang et al., 2022), while rs28362491 is located in the promoter region of the gene, which is crucial in regulating its expression. A polymorphism in this region has the potential to modify transcription factor binding sites, thereby affecting gene expression (Ju et al., 2012). Recent studies have shown that the variant allele containing the ATTG deletion results in reduced levels of the p50 subunit, thus affecting the availability of both the anti-inflammatory p50/p50 homodimer dimer and the pro-inflammatory p50/p65 heterodimer dimer of THE NF-κB pathway (Soydas et al., 2016).

Several clinical and biochemical factors showed a significant association with ASR. The

significant difference in age and BMI (P<0.001 and *P=0.*001 respectively) suggests that increasing BMI and maternal age are associated with the risk of RSA. Our results are consistent with those of the Asian population and different from those observed in the Swedish population, where an increase in BMI is not associated with the risk of the disease (*P=0.*78) (Lo et al., 2012; Ghoshal and Bhattacharyya, 2015).
In addition, lifestyle habits show significant variations, particularly in smoking (*P* = 0.006) and caffeine consumption (*P* = 0.046), which alter placental development. This association has also been confirmed in the Italian population (Stefanidou et al., 2011). Although the use of oral contraceptives is not implicant in the occurrence of the disease (*P* = 0.144).
Biochemically, there were significant differences in LDL (*P* = 0.010) and protein (*P* = 0.001) levels, but not in glucose (*P* = 0.648). These results highlight the importance of taking a multitude of factors into account when assessing the risks associated with reproductive health.
The study of l'genetic association of *NFκBI* gene polymorphisms with l'ASR showed that rs4648143 and rs28362491 are not in HW equilibrium, this could be explained by the origin and current genetic construction of the Tunisian population , which is a heterogeneous population constituted by a тё1апде of different ethnic groups such as Capsians, Phoenicians, Romans, Vandals, Byzantines and Arabs, followed by 1 massive immigration of Bedouins during the 11[eme] century (Hajjej et al., 2006), which makes tracing the genealogical origin of each participant almost impossible. However, re-evaluating ethnic difference is essential for genetic association studies.
In our cohort, rs41275743 is the only one to obey HWE, the frequencies of the minor allele (FAM) are 0.04 in patients and 0.07 in controls. The difference in this allele between the two groups explains the association of this SNP with ASR in our study (***P=0.*013**). Genotypic frequencies showed a significantly high frequency of the heterozygous G/A genotype (1/2) in controls compared with patients, whereas the homozygous A/A mutated genotype (2/2) was not associated with ASR. Indeed, the OR values revealed the association of the G/A genotype as a protective genotype for the risk of RSA in the codominant and dominant modële. However, the G/A and A/A genotypes were associated with an increased risk of acute renal failure' (ARF) in patients with sepsis in the Chinese population (*P*= 0.01; P<0.01 respectively), compared with the G/G genotype, this risk was significantly elevated in the dominant and recessive models (Sun et al., 2020). The disparity in results can be attributed to the distinct origins of the populations studied, and it should be noted that the pathology in question is not identical, making it difficult to establish similar genetic associations between genotypes.
The rs4648143 polymorphism is strongly linked to RSA pathology due to differences in FAM and genotypic frequencies in patients and controls (***P<0.*0001**). Genotypes A/A and G/A are associated with RSA in all three genetic models (codominant, dominant and recessive). Based on the OR values in each genetic modële, rs4648143 is a protective factor, not involved in the occurrence of RSA. Like the preceding polymorphism, the association of rs46418143 with the risk of AKI in patients with sepsis in the Chinese population was significantly higher in the G/A (*P<0.*01) and A/A (*P=0.02)* gënotypes compared with the G/G gënotype in the dominant and recessive modëles. Unlike in the Spanish population, in which no association of this polymorphism with alcoholic liver disease was ële observed (*P=0.*646) (Novo-Veleiro et al., 2018; Sun et al., 2020). The lack of significance obtained in this ëstudy could be explained by the low frequency of the minor al^le observed in this SNP or by the lack of effect of this specific gene variant.

The rs28362491 variant presented l'allële insATTG as the minor al^le, the FAMs as well as the frequencies of the del/ins (1/2), ins/ins (2/2) and del/del (1/1) genotypes showed a significant difference in the two study groups (***P<0.0001***).
Indeed, the allelic distribution of rs28362491 differs significantly between African and European populations, with a disparity of 20%, as well as statistically significant genotypic variations (*P* <0.0001) (Amador et al., 2016). This variant could also be associated with susceptibility to morbid obesity, for which the ins variant (*P=0*.042) and the ins/ins genotype (*P=0*.030) have been associated with increased risk of this disease and increased liver enzyme levels in obese patients (Soydas et al., 2016; Yenmis et al., 2018).
Similarly, the ins allele is implicated in the increased risk of ovarian cancer in the Chinese population due to the central role of the *NFκBI* gene in cell proliferation and apoptosis (Chen et al., 2015). Furthermore, this allele is associated with elevated levels of pro-inflammatory cytokines (TNF-a and IL-6) in type 2 diabetes in the Asian population (Gautam et al., 2017; Chatterjee et al., 2020). This could be due to the active involvement of NF-κB in the generation of inflammation and oxidative stress, triggering the development of diabetic complications.
Also, an Egyptian study on male infertility showed an association of the ins/ins heterozygous genotype with a significantly elevated risk of poor sperm quality in the Egyptian population, in contrast to the Algerian population in which the ins/del heterozygous genotype is associated with an increased risk of chronic inflammatory bowel disease (El-Hoseny et al., 2020; Hamadou et al., 2020).
The rs28362491 polymorphism is associated with an increased risk of RSA in all genetic modëles establishing a 'dose-response' relationship. This result is supported by the haplotypic analysis in which a highly significant association of the TAG (***P=4.5104***$^{e-11}$) and AGG (***P=3.8037***$^{e-7}$) haplotypes were observed for women with ASR, explaining that these haplotypes constitute an important risk factor in the occurrence of this pregnancy complication. The Haploview results also revealed that the TAA haplotype (*P=0.0224)* is associated with the susceptibility of pregnant women to the risk of miscarriage. Thus, the linkage disequilibrium is greater between rs28362491 and rs41275743, to a lesser degree between rs28362491 and rs4648143. However, no binding disequilibrium was shown between rs4648143 and rs41275743.
Molecular genetics is playing an increasingly relevant role in the diagnosis of both monogenic Mendelian and complex multigenic disorders, and offers a promising prospect for the development of new therapies in the human population. The genetics of complex diseases presents enormous difficulties, unlike monogenic disorders, due to their polygenic nature and the great diversity of genetic alterations.
The study of polymorphisms is important as a fundamental element in the study of multifactorial diseases, but it remains highly inadequate for explaining the pathophysiological mechanism of multiple complications such as recurrent abortions. This has prompted researchers to develop new methods for studying genes, such as bioinformatics, which can be used to study the interactions of different genes with other factors influencing their expression, in particular epigenetic factors that modify cell signalling pathways such as NF-κB.
Among the epigenetic factors, the microRNA hsa-miR-21-5p plays a central role in the regulation of inflammation since it affects the expression of the *NFκBI* and *IL-6* genes involved in the immune and inflammatory response.

In our present work, the expression of hsa-miR-21-5 with respect to its target genes, *NFkB1* and *IL-6*, was assessed using real-time PCR. TaqMan technology was used to analyse the specific expression of hsa-miR-21-5p, while SYBR Green technology was used to quantify the expression of the target genes in the patient and the female control.

According to the literature, a study of human placental growth in the Indian population revealed that overexpression of miR-21-5p was negatively associated with overexpression of the *Phosphatase and TENsin homolog* (*PTEN*) gene. Although miR-21-5p could be involved in placental growth, it is unlikely that this regulation is mediated by placental expression of the *PTEN* gene (Kochhar et al., 2022).

According to another study in the Chinese population, the data revealed significant overexpression of miR-21 in macrosomic placental tissue compared with normal control placentas. These findings are also consistent with a previous study in the US population that reported an under-expression of miR-21 expression level in placental tissues of pregnancies with fretal growth retardation (Maccani et al., 2011; Jiang et al., 2014).

Expression of hsa-miR-21-5p is associated with the pathophysiology of pregnancy loss. A study of miR-21-5p expression in women from the Slovakian population suffering from preeclampsia, one of the most severe complications of pregnancy and a cause of fetal death, revealed an overexpression of this microRNA in preeclamptic pregnancies compared with normal pregnancies. However, under-expression of this microRNA has been found in patients with adenomyosis, a condition characterised by the abnormal presence of endometrial tissue in the muscular wall of the uterus and presenting a prevalence of 38.2% in cases of recurrent pregnancy loss (Yan et al., 2019; Bertucci et al., 2023).

Unfortunately, no conclusive conclusions could be drawn from the results obtained, a situation that could be explained by the limitation of using a single sample. This limitation stems from the rarity of finding patients with more than four successive abortions who are still unable to have children, as well as the short duration of the sample. The eventual results of this study could suggest a possible systematic correlation between alteration of the inflammation pathway and spontaneous miscarriage. However, it should be noted that this finding does not constitute a definitive conclusion as to the causality of this complication, given the complexity of RSA due to its multifactorial nature. In this context, it is important to consider this approach as an avenue of scientific research that opens new perspectives on the use of microRNAs as diagnostic and prognostic markers of RSA.

CHAPTER 5

Conclusion and outlook

RSA is a common complication of pregnancy that represents a vëritaЬle concern in human reproduction and can have a ënorrre impact in the life of the couple. A good understanding of the mëcanisms involved in maintaining pregnancy remains a subject of increasing interest to researchers. In fact, monitoring the progress of the pregnancy and detecting any risk of spontaneous abortion can help couples who have had ASR to succeed in having children. Despite scientific advances in this multifactorial disease, it remains complicated and poorly understood.

The association of environmental factors such as maternal age, obesity, smoking and alcoholism with susceptibility genes can induce RSA. Genetic studies are numerous and have contributed to the identification of hundreds of genes involved in RSA, as well as the various polymorphisms whose association with RSA may be determinant or iK'gative. These studies have provided an in-depth understanding of the pathophysiological mechanism of RSA, with a view to developing preventive and curative therapies for this serious complication.

Regarding the gënëtique of ASR, recent ëtudes have rëyëк the involvement of the *NFκBI* gene as a susceptibil^ gene. However, its polymorphisms rs4648143, rs41275743 and in particular rs28362491 INDEL have not ële considered nor ëtudiësed in RSA pathology.

Our ëtude of the three SNPs of the *NFκBI* gene showed a significantly ëlevënotypic gënotypic frequency in patients carrying mutë ins/ins homozygous gënotype of rs28362491 compared with women tëless, Also respectively, homozygous wild-type and hëtërozygous gënotype of rs4648143, homozygous wild-type gënotype of rs41275743. The altalic frequency distribution of rs4648143, rs41275743 and rs28362491 showed a significant difference between patients and control women. Fixing the wild homozygous gënotypes of each variant as rëfërence (OR=1.00), the hëtërozygous and mutës gënotypes of rs4648143 and rs41275743 are consideredërës protective gënotypes against spontaneous abortion, in contrast to those of rs28362491 which are likely to induce RSA. Haplotypic analysis confirmed the results obtained, revealing a significant diffërence of the different haplotypes between patients and tëless women, particularly TAG and AGG which are most associated with women's susceptibility to RSA.

Gënëtic association studies are still essential to better understand the pathophysiology of multifactorial diseases and are the first step in identifying the various biomarkers in order to identify new therapeutic targets and establish a new strategy for managing RSA, such as a gene therapy strategy.

At present, it is clear that the role and potential of microRNAs in the development of RSA is inescapable, since they constitute important diagnostic markers and therapeutic targets, contributing to the development of personalised medicine by providing a molecular definition of the prognosis of RSA. Thus, understanding the regulatory pathways involving microRNAs could help in the treatment of RSA.

Expression analysis of hsa-miR-21-5p and its target genes highlighted a potential link between the inflammation pathway and ASR, opening up prospects for future investigations.

Despite the technical advances in genetics and genomics, other perspectives can be envisaged, in particular the study of the transcriptomics and proteomics of biomarkers identified in aborted tissues. More essentially, new generation sequencing (NGS) technology is a reliable technique for discovering the chromosomal abnormalities and mutations that cause RSA, offering new insights into these pathological mechanisms.

References

A

Account - GeneCards Suite. https://www.genecards.org/cgi-bin/carddisp.pl?gene=NFKB1.

Ali A, Hadlich F, Abbas MW, Iqbal MA, Tesfaye D, Bouma GJ, Winger QA, Ponsuksili S. 2021. MicroRNA-mRNA Networks in Pregnancy Complications:

A

Comprehensive Downstream Analysis of Potential Biomarkers. *Int J Mol Sci* **22**.

Alqudah AM, Sallam A, Stephen Baenziger P, Borner A. 2020. GWAS: Fast-forwarding gene identification and characterization in temperate Cereals: lessons from Barley - A review. *J Adv Res* **22**: 119-135.

Amador MA, Cavalcante GC, Santos NP, Gusmao L, Guerreiro JF, Ribeiro-dos-Santos A, Santos S. 2016. Distribution of allelic and genotypic frequencies of IL1A, IL4, NFKB1 and PAR1 variants in Native American, African, European and Brazilian populations. *BMC Res Notes* **9**: 101.

AquaPortail, Q. (2021) *Small RNA interf rent*.

https://www.aquaportail.com/dictionnaire/definition/14827/petit-arn-interferent.

Arias-Sosa LA, Acosta ID, Lucena-Quevedo E, Moreno-Ortiz H, Esteban-Perez C, Forero-Castro M. 2018. Genetic and epigenetic variations associated with idiopathic recurrent pregnancy loss. *Journal of Assisted Reproduction and Genetics* **35**: 355-366.

Arisawa T, Tahara T, Shiroeda H, Yamada K, Nomura T, Yamada H, Hayashi R, Matsunaga K, Otsuka T, Nakamura M et al. 2013. Functional promoter polymorphisms of NFKB1 influence susceptibility to the diffuse type of gastric cancer. *Oncol Rep* **30**: 3013-3019.

Armistead B, Kadam L, Drewlo S, Kohan-Ghadr HR. 2020. The Role of NFkappaB in Healthy and Preeclamptic Placenta: Trophoblasts in the Spotlight. *Int J Mol Sci* **21**.

ASRM. 2012. Evaluation and treatment of recurrent pregnancy loss - ASRM

Au - Farine T, Au - Parsons M, Au - Lye S, Au - Shynlova O. 2018. Isolation of Primary Human Decidual Cells from the Fetal Membranes of Term Placentae. *JoVE*: e57443.

Bahia W, Soltani I, Abidi A, Haddad A, Ferchichi S, Menif S, Almawi WY. 2020a. Identification of genes and miRNA associated with idiopathic recurrent pregnancy loss: an exploratory data mining study. *BMC Med Genomics* **13**: 75.

Bahia W, Soltani I, Haddad A, Soua A, Radhouani A, Mahdhi A, Ferchichi S. 2020b. Association of genetic variants in Estrogen receptor (ESR)1 and ESR2 with susceptibility to recurrent pregnancy loss in Tunisian women: A case control study. *Gene* **736**: 144406.

Barbaro G, Inversetti A, Cristodoro M, Ticconi C, Scambia G, Di Simone N. 2023. HLA-G and Recurrent Pregnancy Loss. in *International Journal of Molecular Sciences*.

Barrett JC, Fry B, Maller J, Daly MJ. 2004. Haploview: analysis and visualization of LD and haplotype maps. *Bioinformatics* **21**: 263-265.

Bashiri A. 2016. Recurrent Pregnancy Loss.

Baud V, Jacque EJMS-mS. 2008. Alternative NF-κB activation pathway and cancer: Friend or foe? **24**: 1083-1088.

Bayarsaihan D. 2011. Epigenetic mechanisms in inflammation. *J Dent Res* **90**: 9-17.

Benammar A, Sermondade N, Faure C, Dupont C, Cedrin-Durnerin I, Sifer C, Hercberg S, Levy R. 2012. Nutrition and spontaneous miscarriage: a review of the literature. *Gynecologie Obstetrique & Fertilite* **40**: 162-169.

Bender Atik R, Christiansen OB, Elson J, Kolte AM, Lewis S, Middeldorp S, Mcheik S, Peramo B, Quenby S, Nielsen HS et al. 2023. ESHRE guideline: recurrent pregnancy loss: an update in 2022f. *Human Reproduction Open* **2023**.

Bertucci E, Sileo FG, Diamanti M, Alboni C, Facchinetti F, La Marca A. 2023. How adenomyosis changes throughout pregnancy: A retrospective cohort study. **160**: 856-863.

Bianco B, Lerner TG, Trevisan CM, Cavalcanti V, Christofolini DM, Barbosa CP. 2012. The nuclear factor-kB functional promoter polymorphism is associated with endometriosis and infertility. *Hum Immunol* **73**: 1190-1193.

Bigot A. 2009. Identification and study of the expression of detoxification genes in the freshwater bivalves Unio tumidus and Corbicula fluminea: laboratory and natural environment approaches.

B0rsting C, Morling N. 2013. Single-Nucleotide Polymorphisms. in *Encyclopedia of Forensic Sciences (Second Edition)* (eds. JA Siegel, PJ Saukko, MM Houck), pp. 233-238. Academic Press, Waltham.

C

Challis JR, Lockwood CJ, Myatt L, Norman JE, Strauss JF, Petraglia F. 2009. Inflammation and Pregnancy. *Reproductive Sciences* **16**: 206-215.

Chatterjee T, De D, Chowdhury S, Bhattacharyya M. 2020. Nuclear factor NF-kappaB1 functional promoter polymorphism and its expression conferring the risk of Type 2 diabetes-associated dyslipidemia. *Mamm Genome* **31**: 252-262.

Chauhan P, Nair A, Patidar A, Dandapat J, Sarkar A, Saha B. 2021. A primer on cytokines. *Cytokine* **145**: 155458.

Chen L, Deng H, Cui H, Fang J, Zuo Z, Deng J, Li Y, Wang X, Zhao L. 2018. Inflammatory responses and inflammation-associated diseases in organs. *Oncotarget* **9**: 7204-7218.

Chen LP, Cai PS, Liang HB. 2015. Association of the genetic polymorphisms of NFKB1 with susceptibility to ovarian cancer. *Genet Mol Res* **14**: 8273-8282.

Chen X, Guo DY, Yin TL, Yang J. 2021. Non-Coding RNAs Regulate Placental Trophoblast Function and Participate in Recurrent Abortion. *Front Pharmacol* **12**: 646521.

Comba C, Bastu E, Dural O, Yasa C, Keskin G, Ozsurmeli M, Buyru F, Serdaroglu H. 2015. Role of inflammatory mediators in patients with recurrent pregnancy loss. *Fertility and Sterility* **104**: 1467-1474.e1461.

D

Daher S, Denardi KdAG, Blotta MHsSL, Mamoni RL, Reck APM, Camano L, Mattar R. 2004. Cytokines in recurrent pregnancy loss. *Journal of Reproductive Immunology* **62**: 151-157.

Daher S, Mattar R, Gueuvoghlanian-Silva BY, Torloni MR. 2012. Genetic Polymorphisms and Recurrent Spontaneous Abortions: An Overview of Current Knowledge. **67**: 341347.

Dana H, Chalbatani GM, Mahmoodzadeh H, Karimloo R, Rezaiean O, Moradzadeh A, Mehmandoost N, Moazzen F, Mazraeh A, Marmari V et al. 2017. Molecular Mechanisms and Biological Functions of siRNA. *Int J Biomed Sci* **13**: 48-57.

Das K, Rao LVM. 2022. The Role of microRNAs in Inflammation. *Int J Mol Sci* **23**.

Delabaere A, Huchon C, Lavoue V, Lejeune V, Iraola E, Nedellec S, Gallot V, Capmas P, Beucher G, Subtil D et al. 2014. Standardisation of pregnancy loss terminology: expert consensus of the College national des gynecologues et obstëtriciens frangais (CNGOF). *Journal of Obstetric Gynecology and Reproductive* **Biology 43**: 756-763.

Dimitriadis E, Menkhorst E, Saito S, Kutteh WH, Brosens JJ. 2020. Recurrent pregnancy loss. *Nature Reviews Disease Primers* **6**: 98.

Dong AC, Morgan J, Kane M, Stagnaro-Green A, Stephenson MD. 2020. Subclinical hypothyroidism and thyroid autoimmunity in recurrent pregnancy loss: a systematic review and meta-analysis. *Fertility and Sterility* **113**: 587-600.e581.

E

Edenberg HJ, Xuei X, Wetherill LF, Bierut L, Bucholz K, Dick DM, Hesselbrock V, Kuperman S, Porjesz B, Schuckit MA et al. 2008. Association of NFKB1, which encodes a subunit of the transcription factor NF-kappaB, with alcohol dependence. *Hum Mol Genet* **17**: 963-970.

El Hachem H, Crepaux V, May-Panloup P, Descamps P, Legendre G, Bouet PE. 2017. Recurrent pregnancy loss: current perspectives. *Int J Womens Health* **9**: 331-345.

El-Hoseny R, Neamatallah M, Alghobary M, Zalata A, Comhaire F, El-Beah SM. 2020. The possible role of NF-kappaB1 Rs28362491 polymorphism in male fertility of Egyptian population. *Andrologia* **52**: e13659.

ESHRE. 2017. Guideline on the management of recurrent pregnancy loss.

F

Fattal E. 2020. Therapeutic strategies for the clinical application of RNA interferents. *Bulletin de l'Academie Nationale de Medecine* **204**: 1088-1097.

Ford HB, Schust DJ. 2009. Recurrent pregnancy loss: etiology, diagnosis, and therapy. *Rev Obstet Gynecol* **2**: 76-83.

Fu B, Tian Z, Wei H. 2014. TH17 cells in human recurrent pregnancy loss and pre-eclampsia. *Cell Mol Immunol* **11**: 564-570.

G

Gautam A, Gupta S, Mehndiratta M, Sharma M, Singh K, Kalra OP, Agarwal S, Gambhir JK. 2017. Association of NFKB1 gene polymorphism (rs28362491) with levels of inflammatory biomarkers and susceptibility to diabetic nephropathy in Asian Indians. *World J Diabetes* **8**: 66-73.

Ghoshal K, Bhattacharyya M. 2015. Adiponectin: Probe of the molecular paradigm associating diabetes and obesity. *World J Diabetes* **6**: 151-166.

Godines-Enriquez MS, Miranda-Velasquez S, Enriquez-Perez MM, Arce-Sanchez L, Martinez-Cruz N, Flores-Robles CM, Aguayo-Gonzalez P, Morales-Hernandez FV, Villarreal-Barranca A, Suarez-Rico BV et al. 2021. Prevalence of Thyroid Autoimmunity in Women with Recurrent Pregnancy Loss. *Medicina (Kaunas)* **57**.

Gomez-Chavez F, Correa D, Navarrete-Meneses P, Cancino-Diaz JC, Cancino-Diaz ME, Rodriguez-Martmez S. 2021. NF-κB and Its Regulators During Pregnancy. **12**.

Graham JJ, Longhi MS, Heneghan MA. 2021. T helper cell immunity in pregnancy and influence on autoimmune disease progression. *J Autoimmun* **121**: 102651.

H

Hajjej A, Kaabi H, Sellami MH, Dridi A, Jeridi A, El borgi W, Cherif G, Elgaaied A, Almawi WY, Boukef K et al. 2006. The contribution of HLA class I and II alleles and haplotypes to the investigation of the evolutionary history of Tunisians. **68**: 153162.

Hakimi P, Tabatabaei F, Rahmani V, Zakariya NA, Moslehian MS, Bedate AM, Tamadon A, Rahbarghazi R, Mahdipour M. 2023. Dysregulated miRNAs in recurrent miscarriage: A systematic review. *Gene* **884**: 147689.

Hamadou I, Garritano S, Romanel A, Naimi D, Hammada T, Demichelis F. 2020. Inherited variant in NFkappaB-1 promoter is associated with increased risk of IBD in an Algerian population and modulates SOX9 binding. *Cancer Rep (Hoboken)* **3**: e1240.

Hamilton JP. 2011. Epigenetics: principles and practice. *Dig Dis* **29**: 130-135.

Hocher B, Hocher CF. 2018. Epigenetics of recurrent pregnancy loss. *EBioMedicine* **35**: 1819.

Hoeger B, Serwas NK, Boztug K. 2017. Human NF-kappaB1 Haploinsufficiency and Epstein- Barr Virus-Induced Disease-Molecular Mechanisms and Consequences. *Front Immunol* **8**: 1978.

Huxford T, Ghosh G. 2009. A structural guide to proteins of the NF-kappaB signaling module. *Cold Spring Harb Perspect Biol* **1**: a000075.

Hyde KJ, Schust DJ. 2015. Genetic considerations in recurrent pregnancy loss. *Cold Spring Harb Perspect Med* **5**: a023119.

Jairajpuri DS, Malalla ZH, Mahmood N, Khan F, Almawi WY. 2021. Differentially expressed circulating microRNAs associated with idiopathic recurrent pregnancy loss. *Gene* **768**: 145334.

Jauniaux E, Farquharson RG, Christiansen OB, Exalto N. 2006. Evidence-based guidelines for the investigation and medical treatment of recurrent miscarriage. *Hum Reprod* **21**: 2216-2222.

Jiang H, Wu W, Zhang M, Li J, Peng Y, Miao Tt, Zhu H, Xu G. 2014. Aberrant upregulation of miR-21 in placental tissues of macrosomia. *Journal of Perinatology* **34**: 658-663.

Jiang S, He F, Gao R, Chen C, Zhong X, Li X, Lin S, Xu W, Qin L, Zhao X. 2021. Neutrophil and Neutrophil-to-Lymphocyte Ratio as Clinically Predictive Risk Markers for Recurrent Pregnancy Loss. *Reproductive Sciences* **28**: 1101-1111.

Jie M, Feng T, Huang W, Zhang M, Feng Y, Jiang H, Wen Z. 2021. Subcellular Localization of miRNAs and Implications in Cellular Homeostasis. *Genes (Basel)* **12**.

Ju Z, Zheng X, Huang J, Qi C, Zhang Y, Li J, Zhong J, Wang C. 2012. Functional characterization of genetic polymorphisms identified in the promoter region of the bovine PEPS gene. *DNA Cell Biol* **31**: 1038-1045.

K

Kanigur sultuybek G, Yenmis G, Soydas T. 2020. The Functional Variations of NFKB1 and NFKB1A in inflammatory disorders and Therapeutic Approaches. *Asian biomedicine* **14**.

Karban AS, Okazaki T, Panhuysen CIM, Gallegos T, Potter JJ, Bailey-Wilson JE, Silverberg MS, Duerr RH, Cho JH, Gregersen PK et al. 2004. Functional annotation of a novel NFKB1 promoter polymorphism that increases risk for ulcerative colitis. *Human Molecular Genetics* **13**: 35-45.

Kaser D. 2018. The Status of Genetic Screening in Recurrent Pregnancy Loss. *Obstet Gynecol Clin North Am* **45**: 143-154.

Kazerooni T, Ghaffarpasand F, Asadi N, Dehkhoda Z, Dehghankhalili M, Kazerooni Y. 2013. Correlation between thrombophilia and recurrent pregnancy loss in patients with polycystic ovary syndrome: A comparative study. *Journal of the Chinese Medical Association* **76**: 282-288.

Kiselev IS, Kulakova OG, Boyko AN, Favorova OO. 2021. DNA Methylation As an Epigenetic Mechanism in the Development of Multiple Sclerosis. *Acta Naturae* **13**: 45-57.

Kochhar P, Dwarkanath P, Ravikumar G, Thomas A, Crasta J, Thomas T, Kurpad AV, Mukhopadhyay A. 2022. Placental expression of miR-21-5p, miR-210-3p and miR- 141-3p:

relation to human fetoplacental growth. *European Journal of Clinical Nutrition* **76**: 730-738.

Korzeniewski S, Hofman P, Brest PJMS. 2013. Silent rather than noisy polymorphisms. **29**: 124-126.

Kutteh WH, Stephenson MD. 2006. Chapter 55 - Recurrent Pregnancy Loss. in *Clinical Gynecology* (eds. EJ Bieber, JS Sanfilippo, IR Horowitz), pp. 797-802, Churchill Livingstone, Philadelphia.

Kwak-Kim J, Bao S, Lee SK, Kim JW, Gilman-Sachs A. 2014. Immunological Modes of Pregnancy Loss: Inflammation, Immune Effectors, and Stress. *American Journal of Reproductive Immunology* **72**: 129-140.

L

Lee HL, Jang JW, Lee SW, Yoo SH, Kwon JH, Nam SW, Bae SH, Choi JY, Han NI, Yoon SK. 2019. Inflammatory cytokines and change of Th1/Th2 balance as prognostic indicators for hepatocellular carcinoma in patients treated with transarterial chemoembolization. *Scientific Reports* **9**: 3260.

Lëgarë C, Cement A-A, Desgag^ V, Thibeault K, White F, Guay S-P, Arsenault BJ, Scott MS, Jacques P-E, Perron P et al. 2022. Human plasma pregnancy-associated miRNAs and their temporal variation within the first trimester of pregnancy. *Reproductive Biology and Endocrinology* **20**: 14.

Liu T, Zhang L, Joo D, Sun S-C. 2017. NF-κB signaling in inflammation. *Signal Transduction and Targeted Therapy* **2**: 17023.

Lo W, Rai R, Hameed A, Brailsford SR, Al-Ghamdi AA, Regan L. 2012. The effect of body mass index on the outcome of pregnancy in women with recurrent miscarriage. **19**: 167-171.

Logan MK, Lett KE, McLaurin DM, Hebert MD. 2022. Coilin as a regulator of NF-kB mediated inflammation in preeclampsia. *Biol Open* **11**.

Lordan R, Tsoupras A, Zabetakis I. 2019. Chapter 2 - Inflammation. in *The Impact of Nutrition and Statins on Cardiovascular Diseases* (eds. I Zabetakis, R Lordan, A Tsoupras), pp. 23-51. Academic Press.

M

Maccani MA, Padbury JF, Marsit CJ. 2011. miR-16 and miR-21 expression in the placenta is associated with fetal growth. *PLoS One* **6**: e21210.

Mattick JS, Amaral PP, Carninci P, Carpenter S, Chang HY, Chen L-L, Chen R, Dean C, Dinger ME, Fitzgerald KA et al. 2023. Long non-coding RNAs: definitions, functions, challenges and recommendations. *Nature Reviews Molecular Cell Biology* **24**: 430-447.

Merviel P, Lanta S, Allier G, Gagneur O, Najas S, Nasreddine A, Campy H, Verhoest P, Naepels P, Gondry J et al. 2005. Spontaneous abortions a repetition. *EMC - Gynecologie-Obstetrique* **2**: 278-296.

Miller, S.A., Dykes, D.D. and Polesky, H.F. 1988. A simple salting out procedure for extracting DNA from human nucleated cells. *Nucleic Acids Research*, **16**: 1215-1215.

Misra MK, Singh B, Mishra A, Agrawal S. 2016. Co-stimulatory CD28 and transcription factor NFKB1 gene variants affect idiopathic recurrent miscarriages. *Journal of Human Genetics* **61**: 1035-1041.

Montpetit A, Chagnon FJMS. 2006. The Haplotype Map of the human дёпоте. **22**: 1061-1068.

Mor G, Cardenas I, Abrahams V, Guller S. 2011. Inflammation and pregnancy: the role of the immune system at the implantation site. *Ann N Y Acad Sci* **1221**: 80-87.

Mtiraoui N, Borgi L, Hizem S, Nsiri B, Finan RR, Gris J-C, Almawi WY, Mahjoub T. 2005. Prevalence of antiphospholipid antibodies, factor V G1691A (Leiden) and prothrombin G20210A mutations in early and late recurrent pregnancy loss. *European Journal of Obstetrics & Gynecology and Reproductive Biology* **119**: 164170.

N

Neininger K, Marschall T, Helms V. 2019. SNP and indel frequencies at transcription start sites and at canonical and alternative translation initiation sites in the human genome. *PLoS One* **14**: e0214816.

Ng KYB, Cherian G, Kermack AJ, Bailey S, Macklon N, Sunkara SK, Cheong Y. 2021. Systematic review and meta-analysis of female lifestyle factors and risk of recurrent pregnancy loss. *Sci Rep* **11**: 7081.

Novo-Veleiro I, Cieza-Borrella C, Pastor I, Gonzalez-Sarmiento R, Laso FJ, Marcos M. 2018. Analysis of the relationship between interleukin polymorphisms within miRNA- binding regions and alcoholic liver disease. *Rev Clin Esp (Barc)* **218**: 170-176.

O

O'Brien J, Hayder H, Zayed Y, Peng C. 2018. Overview of MicroRNA Biogenesis, Mechanisms of Actions, and Circulation. *Front Endocrinol (Lausanne)* **9**: 402.

P

Park JC, Han JW, Lee SK. 2022. Chapter 3 - T helper cell pathology and recurrent pregnancy losses; Th1/Th2, Treg/Th17, and other T cell responses. in *Immunology of Recurrent Pregnancy Loss and Implantation Failure* (ed. J Kwak-Kim), pp. 27-53. Academic Press.

Pei C-Z, Kim YJ, Baek K-H. 2019. Pathogenetic factors involved in recurrent pregnancy loss from multiple aspects. *ogs* **62**: 212-223.

Peterson JM, Bakkar N, Guttridge DC. 2011. Chapter four - NF-κB Signaling in Skeletal Muscle Health and Disease. in *Current Topics in Developmental Biology* (ed. Gk Pavlath), pp. 85-119. Academic Press.

Priya PK, Mishra VV, Roy P, Patel H. 2018. A Study on Balanced Chromosomal Translocations in Couples with Recurrent Pregnancy Loss. *J Hum Reprod Sci* **11**: 337-342.

R

RNACentral (no date). https://rnacentral.org/rna/URS000039ED8D/9606.

RS41275743 REFSNP Report - DBSNP - NCBI (2022). https://www.ncbi.nlm.nih.gov/snp/rs41275743.

RS4648143 REFSNP Report - DBSNP - NCBI (2022). https://www.ncbi.nlm.nih.gov/snp/rs4648143.

S

Salimi S, Sargazi S, Mollashahi B, Heidari Nia M, Mirinejad S, Majidpour M, Ghasemi M, Sargazi S. 2022. Association of Polymorphisms in miR146a, an Inflammation- Associated MicroRNA, with the Risk of Idiopathic Recurrent Spontaneous Miscarriage: A Case-Control Study. *Dis Markers* **2022**: 1495082.

Savagner F, Le Pennec S, Rivalin R, Eyer J. 2015. Les microARNs. *Revue Francophone des Laboratoires* **2015**: 49-54.

Sejoume N, Callahan S, Chabrol H. 2008. The psychological impact of miscarriage: a review of the literature. *Journal de Gynecologie Obstetrique et Biologie de la Reproduction* **37**: 435-

440.
Shaheen, Saxena U. 2022. Association between Chlamydia trachomatis infection and recurrent pregnancy loss. *International Journal of Reproduction, Contraception, Obstetrics and Gynecology* **11**: 557+.
Soydas T, Karaman O, Arkan H, Yenmis G, Ilhan MM, Tombulturk K, Tasan E, Kanigur Sultuybek G. 2016. The Correlation of Increased CRP Levels with NFKB1 and TLR2 Polymorphisms in the Case of Morbid Obesity. *Scand J Immunol* **84**: 278283.
Statello L, Guo C-J, Chen L-L, Huarte M. 2021. Gene regulation by long non-coding RNAs and its biological functions. *Nature Reviews Molecular Cell Biology* **22**: 96-118.
Stefanidou EM, Caramellino L, Patriarca A, Menato G. 2011. Maternal caffeine consumption and sine causa recurrent miscarriage. *European Journal of Obstetrics & Gynecology and Reproductive Biology* **158**: 220-224.
Sun J, Cai X, Shen J, Jin G, Xie Q. 2020. Correlation Between Single Nucleotide Polymorphisms at the 3'-UTR of the NFKB1 Gene and Acute Kidney Injury in Sepsis. *Genet Test Mol Biomarkers* **24**: 274-284.
Sun SC. 2017. The non-canonical NF-kappaB pathway in immunity and inflammation. *Nat Rev Immunol* **17**: 545-558.

T

Toupet A, Tlieau A, Goffinet F, Tsatsaris V. 2015. Pregnancy losses a гёрёййоп: ёtiologies and workup, the gynёcologist-obstётrician perspective. *La Revue de Medecine Interne* **36**: 182-190.
Treiber T, Treiber N, Meister G. 2019. Regulation of microRNA biogenesis and its crosstalk with other cellular pathways. *Nature Reviews Molecular Cell Biology* **20**: 5-20.
Turesheva A, Aimagambetova G, Ukybassova T, Marat A, Kanabekova P, Kaldygulova L, Amanzholkyzy A, Ryzhkova S, Nogay A, Khamidullina Z et al. 2023. Recurrent Pregnancy Loss Etiology, Risk Factors, Diagnosis, and Management. Fresh Look into a Full Box. *J Clin Med* **12**.
Turocy JM, Rackow BW. 2019. Uterine factor in recurrent pregnancy loss. *Semin Perinatol* **43**: 74-79.

U

Uffelmann E, Huang QQ, Munung NS, de Vries J, Okada Y, Martin AR, Martin HC, Lappalainen T, Posthuma D. 2021. Genome-wide association studies. *Nature Reviews Methods Primers* **1**: 59.

V

Vieira ML, Santini L, Diniz AL, Munhoz Cde F. 2016. Microsatellite markers: what they mean and why they are so useful. *Genet Mol Biol* **39**: 312-328.
Vomstein K, Feil K, Strobel L, Aulitzky A, Hofer-Tollinger S, Kuon RJ, Toth B. 2021. Immunological Risk Factors in Recurrent Pregnancy Loss: Guidelines Versus Current State of the Art. *J Clin Med* **10**.

W

Wang W, Zhao Y, Zhou X, Sung N, Chen L, Zhang X, Ma D, Zhu P, Kwak-Kim J. 2022. Dynamic changes in regulatory T cells during normal pregnancy, recurrent pregnancy loss, and gestational diabetes. *Journal of Reproductive Immunology* **150**: 103492.
Witte, J.S. 2010. Genome-Wide Association Studies and Beyond. *Annual Review of Public*

Health, **31**: 9-20.

X

Xu N, Zhou X, Shi W, Ye M, Cao X, Chen S, Xu C. 2022. Integrative analysis of circulating microRNAs and the placental transcriptome in recurrent pregnancy loss. **13**.

Y

Yan Q, Yan G, Zhang C, Wang Z, Huang C, Wang J, Zhou J, Liu Y, Ding L, Zhang Q et al. 2019. miR-21 reverses impaired decidualization through modulation of KLF12 and NR4A1 expression in human endometrial stromal cellsf. *Biology of Reproduction* **100**: 1395-1405.

Yang L, Cai C, Fang S, Hao S, Zhang T, Zhang L. 2022. Changes in expression of nuclear factor kappa B subunits in the ovine thymus during early pregnancy. *Scientific Reports* **12**: 17683.

Yang W, Kang X, Yang Q, Lin Y, Fang M. 2013. Review on the development of genotyping methods for assessing farm animal diversity. *Journal of Animal Science and Biotechnology* **4**: 2.

Yang X, Tian Y, Zheng L, Luu T, Kwak-Kim J. 2023. The Update Immune-Regulatory Role of Pro- and Anti-Inflammatory Cytokines in Recurrent Pregnancy Losses. **24**: 132.

Yazdani N, Shekari Khaniani M, Bastami M, Ghasemnejad T, Afkhami F, Mansoori Derakhshan S. 2018. HLA-G regulatory variants and haplotypes with susceptibility to recurrent pregnancy loss. *International Journal of Immunogenetics* **45**: 181-189.

Yenmis G, Soydas T, Arkan H, Tasan E, Kanigur Sultuybek G. 2018. Genetic Variation in NFKB1 Gene Influences Liver Enzyme Levels in Morbidly Obese Women. *Arch Iran Med* **21**: 13-18.

Yu W, Bao S. 2022. Association of male factors with recurrent pregnancy loss. *Journal of Reproductive Immunology* **154**: 103758.

Z

Zhang JM, An J. 2007. Cytokines, inflammation, and pain. *Int Anesthesiol Clin* **45**: 27-37.

Zhang N, Huang D, Jiang G, Chen S, Ruan X, Chen H, Huang J, Liu A, Zhang W, Lin X et al.

2022. Genome-Wide 3'-UTR Single Nucleotide Polymorphism Association Study Identifies Significant Prostate Cancer Risk-Associated Functional Loci at 8p21.2 in Chinese Population. **9**: 2201420.

Zhang Q, Tian P, Xu H. 2021. MicroRNA-155-5p regulates survival of human decidua stromal cells through NF-κB in recurrent miscarriage. *Reproductive Biology* **21**: 100510.

Zhao Y, Zhang X, Du N, sun H, Chen L, Bao H, Zhao Q, Qu Q, Ma D, Kwak-Kim J et al. 2020. Immune checkpoint molecules on T cell subsets of pregnancies with preeclampsia and gestational diabetes mellitus. *Journal of Reproductive Immunology* **142**: 103208.

Zhou Q, Xiong Y, Qu B, Bao A, Zhang Y. 2021. DNA Methylation and Recurrent Pregnancy Loss: A Mysterious Compass? *Front Immunol* **12**: 738962.

Zhu X, Chen Z, Shen W, Huang G, Sedivy JM, Wang H, Ju Z. 2021. Inflammation, epigenetics, and metabolism converge to cell senescence and ageing: the regulation and intervention. *Signal Transduction and Targeted Therapy* **6**: 245.

MIX
Papier aus verantwortungsvollen Quellen
Paper from responsible sources
FSC® C105338

Printed by Books on Demand GmbH, Norderstedt / Germany